Molecular Interpretations of Sorption in Polymers

Part I

By L. A. Errede

With 64 Figures

Springer-Verlag Berlin
Heidelberg GmbH

Louis A. Errede

3M Corporate Research Laboratories,
3M Center, Bldg 201-2N-22,
St. Paul, MN 55133-3221, USA

ISBN 978-3-662-15001-6   ISBN 978-3-540-46323-8 (eBook)
DOI 10.1007/978-3-540-46323-8

Library of Congress Catalog Card Number 61-642

© Springer-Verlag Berlin Heidelberg 1991

Originally published by Springer-Verlag Berlin Heidelberg New York in 1991.
Softcover reprint of the hardcover 1st edition 1991

Typesetting: Th. Müntzer, Bad Langensalza; Printing: Heenemann, Berlin;

2152/3020-543210 — Printed on acid-free paper

Editors

Table of Contents

The major part of this review is in fact a progress report of an ongoing research effort at 3M Laboratories aimed at a better understanding of polymer swelling by liquids. Since this effort builds naturally upon the many contributions of earlier investigators, Section 1 and 2 summarize the more important of these contributions. A sincere effort is made, however, to provide a good if not complete list of references to the original literature, and to several excellent in-depth reviews that describe these contributions in much more detail. Section 3 describes 3M contributions that lead to the concept of an adsorption parameter, α, the number of adsorbed molecules per monomer unit of polymer at liquid-saturation, which is calculated from the observed relative swelling power of a test-liquid for the sorbent polymer. The accumulation of data obtained in these investigations show that α (with respect to polystyrene) is consistent with expectation based on the molecular structures of the sorbed liquid and the monomer unit of sorbent polymer. Section 4 shows how α relates to the Hildebrand Solubility Parameter, δ, and the Flory-Huggins Interaction Parameter, χ, using δ and χ as reported in the literature and α for the corresponding polystyrene-liquid system as reported in Sect. 3. The relationships established thereby were used in turn to calculate values for δ and χ for polystyrene-liquid systems not yet reported in the literature. Section 4 also reports how α was used to gain better insight into the phenomenon of gel-formation from solutions of isotactic polystyrene in various liquids. Section 5 suggests ways in which might help in the understanding of other phenomena in which molecular adsorption plays an important role, such as polymer permeation, separations by polymer membranes, chromatography, catalysis, and perhaps even the kinetics and selectivity of organic reactions that occur in solution.

1 Introduction

The phenomenon of polymer swelling, owing to sorption of small molecules, was known even before Staudinger reported [1] in 1935 that crosslinked poly(styrene) swells enormously in certain liquids to form two-component polymer gels. The physical state of such systems varies with the concentration (C) and molecular structure of the sorbed molecules: thus, the system undergoes transition at constant temperature from a rigid state (glassy or partially crystalline) at $C < C_g$ to a rubbery state at C_g (the transition state composition). When $C > C_g$ and the second component is a liquid, its subsequent sorption proceeds quickly to gel-saturation; and of course a solution is produced if the polymer lacks covalently bonded crosslinks or equivalent restraints. Each successive physical state exhibits its own characteristic sorption isotherm and sorption kinetics.

So far most of the pertinent literature interprets interactions between polymer and a second sorbed component on the basis of the colligative properties of the system. This information is summarized in Sect. 2 of this review. From those reports, I inferred that some sort of association of the small penetrant molecules with the repeat unit of the polymer is a prerequisite condition for the sequential changes described above. This implies that the magnitude of polymer swelling should be directly related to how well the molecular structure of the small molecule can be accommodated by the molecular structure of the polymer repeat unit in view of the macrostructure of that polymer (i.e. the crosslink density). The results obtained in our laboratory, which are summarized in Sect. 3 of this Review, are thoroughly consistent with that hypothesis. In Sect. 4 these interpretations are extended to correlate the colligative properties of Sect. 2.

2 Background: Colligative-Property Interpretations

2.1 Sorption and Diffusion in Polymeric Systems

2.1.1 Before Transition to the Rubbery State

The economic importance of polymers as physical barriers to fluids in the packaging industry and in separation processes is well known [2, 3]. Consequently sorption of small molecules by a rigid polymer, i.e. in the glassy or crystalline state, has been studied more extensively than has the subsequent sorption after the transition to the rubbery state. It was pointed out by Meares [4] and later by Barrer [5] that sorption of gases and vapors by a polymer in the glassy state ($C < C_g$) is represented best by a dual-mode process [6], i.e. the observed sorption isotherm is a combination of a Henry's law "dissolved" component term, C_D, and a Langmuir "hole filling" term, C_H, the latter pertaining to the so-called free-volume in the rigid state of polymers. The relations are:

$$C = C_D + C_H, \qquad (1)$$

$$C = k_D + C'_H bP/(1 + bP), \qquad (2)$$

where k_D is the Henry's law constant, P is the partial pressure of the sorbed component, and C′ and b are the Langmuir capacity constant and affinity constant respectively. As the concentration, C, of sorbed material approaches C_g, the concentration at which the system changes from the glassy to the rubbery state, the dual-mode sorption isotherm passes through a positive inflection point and thereafter follows the Flory-Huggins mode [7], owing to plasticization of the system by the second component.

So long as the concentration of penetrant molecules at the surface of the polymer in contact with penetrant in a fluid state is less than C_g, the flux, F, of these molecules [down the concentration gradient (dC/dx) in the direction normal to and away from the interface] is given by Fick's law:

$$F = DS(dC/dx),\tag{3}$$

where D and S are the diffusivity and solubility respectively of the penetrant molecules in the "rigid" matrix.

Berens [7] reported that under such conditions the logarithm of diffusivity is a linear function of the molecular diameter of the penetrant molecule. This is consistent with the hypothesis discussed in detail by Stern and Frisch [8, 9] that diffusion through a rigid polymeric matrix is proportional to the energy required to expand (i.e. swell temporarily on a molecular scale) the polymeric chains sufficiently to allow "peristaltic" motion of the penetrant molecules along these chains. This energy is presumed to be proportional to the molecular diameter of the penetrant.

2.1.2 After Transition to the Rubbery State

When the concentration of sorbed molecules at the interface of polymer and fluid penetrant is greater than C_g, plasticization occurs at this interface; and the domain of rubbery-state polymer progresses in step with the permeation front, thereby complicating the kinetics of diffusion accordingly. In the case that the contacting fluid is a liquid, the rubbery state is followed by rapid swelling to saturation such that the front of "swelled-to-saturation" polymer trails closely behind the diffusion front, the progress of which is diffusion-controlled in accordance with Fick's law (Eq. 3). The concentration gradient between the two fronts remains almost constant and very steep. The overall kinetics of diffusion, therefore, is modified Fickian, i.e. Case II diffusion [10], which is typical for glassy polymers, such as poly-(methacrylates) and poly(styrenes), immersed in swelling liquids [11−13].

The physics of sorption and diffusion through polymeric materials in the rigid and rubbery states is discussed thoroughly in several outstanding reviews [14−18].

2.1.3 Sorption of Liquid to Attain Gel-Saturation

In the case that the fluid in contact with the polymeric surface is a liquid, the amount of that liquid sorbed to attain gel-saturation (or solution) after C becomes C_g is of course considerably greater than the amount sorbed to attain C_g. The kinetics of sorption thereafter, however, is dependent upon the history of the polymeric sample [16], and therefore difficult to interpret. This is especially true in the case of water-swellable polymers such as poly(acrylamide) [19], because trace impurities sorbed on the surface affect wetting characteristics and therefore affect the rate of water uptake accordingly.

Although the kinetics of liquid uptake to attain gel-saturation is history-dependent, the composition at the true end-state (i.e. thermodynamic equilibrium in excess liquid) is not; therefore the observed end-state is usually reproducible [19]. Gel-saturation is attained when the restraining force (per unit area) of the polymeric crosslinked network becomes equal and opposite to the osmotic pressure that causes the system to swell [20]. In other words saturation is achieved when the chemical potential of swelling liquid, μ_1, in the swollen network is equal to the chemical potential of the excess pure liquid, μ_1^0, outside the network. It was logical to anticipate that the volume of liquid sorbed per gram of polymer, at this state of thermodynamic equilibrium with excess liquid, would correlate with the molecular structure of the liquid. In fact two parameters already exist which relate the sorption affinity to the molecular structure, namely the solubility parameter, δ, first proposed by Hildebrand [21], and the interaction parameter, χ, introduced by Flory [22] and Huggins [23–26].

2.2 Parameters that Relate to Swelling Power

2.2.1 Hildebrand Solubility Parameter

Hildebrand and Scott [21] have shown that the solubility of a non-electrolyte in non-ionic solvents depends on the similarity of the thermodynamic properties of solute and solvent. They pointed out that *when entropic effects can be ignored*, the free energy of mixing of solute in solvent is determined by the sign and the magnitude of the enthalpy (ΔH). According to Hildebrand, ΔH per unit volume for such endothermic mixing is given by:

$$\Delta H = \phi_1 \phi_2 (\delta_1 - \delta_2)^2 , \tag{4}$$

where ϕ_1 and ϕ_2 are the volume fractions of the solvent and solute respectively, and $(\delta_1 - \delta_2)$ is the difference in solubility parameters. Hildebrand defined δ to be the square root of the cohesive energy density (δ^2) of the molecular species.

$$\delta^2 = \Delta E/V = (\Delta H - P \, \Delta V)/V , \tag{5}$$

where ΔE is the internal energy, which is equal to the enthalpy when $P \, \Delta V$ is zero, and V is the molar volume.

Thermodynamic relationships stated in Eqs. 4 and 5 imply that the more equal the cohesive energy densities of solute and solvent the greater is their mutual compatibility [21] (i.e. like dissolves like).

Marked differences between observed and predicted results occur, however, when the entropy effects are too great to be ignored. In the derivation of Eqs. 4 and 5, only dispersion forces between molecular components were taken into account. The cohesive energy, however, is also dependent upon associations between polar groups and especially hydrogen bonding. It was suggested [27, 28], therefore, that the "total solubility parameter" (δ_t) is represented better by the combined effect of three components.

$$\delta_t^2 = \delta_d^2 + \delta_p^2 + \delta_h^2, \tag{6}$$

where δ_d refers to the contribution by non-polar London dispersion forces (as considered by Hildebrand), δ_p refers to the combined Keesom dipole-dipole forces and Debye dipole-induced-dipole forces, and δ_h refers to Lewis acid-base or hydrogen-bonding forces. Fowkes [29] pointed out that δ_h for polymer-liquid systems can be calculated accurately from the acid-base relationships established by Drago [30, 31].

The solubility parameters of many volatile liquids have been calculated directly from their respective heats of vaporization and molar volumes (Eq. 5). Hoy [32] has shown that δ for relatively non-volatile liquids can be calculated from vapor pressure data using a modification of the Haggenmacher Eq. [33]. Large numbers of such data have been reported and these are collected in extensive tables [27, 28, 34].

Since the cohesive energy density (Eq. 5) is a function of the molecular structure, it was logical to attempt establishment of solvent power for a given liquid as an additive function of the molecular component contributions to the thermodynamic properties. Scatchard [35] has shown that in a homologous series of the type $R(CH_2)_nH$ [where R is H, X, Ph, OH, or OR], $(EV)^{1/2}$ is a linear function of n. Small [36] applied this observation to find additive constants for the more common groups in organic chemistry, which he termed molar attractive constants, i.e. $F_i = (E_iV_i)^{1/2}$. When the F_i constants are summed over the groups present in the molecule, the $(EV)^{1/2}$ is obtained for one mole of the substance, i.e. $\Sigma F_i = (EV)^{1/2}$; the molar cohesive energy, E, cohesive energy density, δ^2, and solubility parameter are then given by:

$$E = (\Sigma F)^2/V; \qquad \delta^2 = (\Sigma F/V)^2; \quad \text{and} \quad \delta = \Sigma F/V.$$

Small [36] showed that most of the reported δ can be estimated by the last of the above three relationships, on the basis of the molecular structure of the solvent. He used this method to calculate the solubility parameter of poly(styrene) to obtain a value, $\delta_{pol} = 9.12$ $(cal/mL)^{1/2}$, which is within the range of the experimental δ-values [8.6 to 9.7 $(cal/mL)^{1/2}$] reported for this polymer [34, 37–39], [note that $(joule/mL)^{1/2} = 2.046$ $(cal/mL)^{1/2}$].

Small's success in calculating δ_{pol} for poly(styrene) encouraged others to attempt calculation of the corresponding δ_{pol} for a variety of other polymers and copolymers by means of component contributions. The value of such calculations are suspect, however, because the component contributions used to calculate the corresponding δ_{pol} were deduced from cohesive energy density data for small molecules, for which the entropy effects are not very significant; this is not the case, however, for high molecular weight multifunctional molecules, especially polymers.

The classical method for measuring the solubility parameter of a polymer is to determine the solubility of that polymer in a large number of liquids with known solubility parameters, and then to plot the observed solubility as a function of the solubility parameters of the test liquid. The solubility parameter where the polymer solubility is maximal [40] is by definition the solubility parameter of the polymer, δ_{pol}.

Gee [41] has shown that the same is normally true for swellability of crosslinked polymers, i.e. the magnitude of swelling exhibited by a polymer with δ_{pol} in a liquid with δ_{liq} increases as ($\delta_{pol} - \delta_{liq}$) approaches zero. Such correlations have been used to establish the solubility parameters of many polymers and these data are also collected in extensive tables [27, 28].

2.2.2 Flory-Huggins Interaction Parameter, χ

The Flory-Huggins theory [20, 22–26, 42, 43] considers the change in potential energy in going from the pure states of polymer and liquid to a mixture thereof. In the original form of this theory, the interaction parameter, χ, between liquid and polymer was defined as

$$\chi = z\,\Delta w_h/kT, \tag{7}$$

where k is the Boltzmann constant, z is the "number of nearest neighbors", Δw_h is the exchange interaction energy defined as the difference between the intermolecular potential, w_{12}, for a molecule of solvent interacting with a neighboring segment of polymer and the mean of the energy of interaction, w_{11}, between two molecules of solvent and the energy of interaction, w_{22}, between two segments of polymer, i.e.:

$$\Delta w_h = w_{12} - 0.5\,(w_{11} + w_{22}). \tag{8}$$

Originally χ was stated to be independent of polymer concentration. The χ-parameters determined by many investigators using one or another of the methods for measuring colligative properties of polymer-liquid solutions (mentioned below) show that this is not the case (see Tables 3–22 of Reference 43); nor does χ vary linearly with $1/T$ as stated in Eq. 7. Later [44] a quantity Δw_s representing an entropic contribution from contact interaction was added to the Flory-Huggins definition of χ to produce a relationship linear in $1/T$.

$$\chi = z(\Delta w_h - w_s)/kT. \tag{9}$$

Despite the above shortcomings, the Flory-Huggins theory has served as a useful model for predicting a large number of colligative properties, and has had remarkable success in providing qualitative if not quantitative descriptions of polymer-solvent systems.

Flory and Huggins have shown that the magnitude of association between polymer and sorbed liquid is reflected in the colligative physical properties of the resultant mixtures. In such cases, χ is given by:

$$\chi = (\mu_1 - \mu_1^0)/RT\phi_2^2 - [\ln(1 - \phi_2) + \phi_2(1 - V_1^0/V_2^0)]/\phi_2^2, \quad (10)$$

where μ_1 and μ_1^0 are the chemical potentials of the liquid in solution and in pure state respectively, R is the gas constant, T is temperature, ϕ_2 is the molar fraction of the polymer and V_1^0 and V_2^0 are the volume fractions of the liquid and polymer respectively. The interaction parameter has been used as a semiquantitative measure of the "goodness" of a solvent in a given polymer-solvent system [42, 43]. A value of $\chi < 0.4$ indicates that the liquid has a relatively good affinity for the polymer, whereas a value of $\chi > 0.8$ indicates the opposite.

The χ-parameter has been related to gas sorption phenomena in polymers using a variation of Eq. 10 obtained by substituting $RT \ln (P/P_0)$ for $(\mu_1 - \mu_1^0)$ to give:

$$\chi = [\ln P_1/(1 - \phi_2) P_1^0] - \phi_2(1 - V_1^0/V_2^0)]/\phi_2^2 \quad (11)$$

which enables one to establish χ from the sorption isotherm [44], or conversely to use a known χ to charcterize the sorption isotherm exhibited by sorbed components that cause plasticization of glassy polymers [45].

Other relationships between χ and an observable physical property such as osmotic pressure [20, 43], freezing point depression of polymer [20, 52] or solvent [20, 53], and gas liquid chromatography [46–54], were established in like fashion. The relationship determined for swelling of cross-linked polymer to thermo-dynamic equilibrium in excess liquid has particular significance for the subject of this review. It is given here in the form of the Flory-Rehner equation.

$$\chi = [V_1^0(v_e/V_2^0) (\phi_2/2 - \phi_2^{1/3}) - \ln(1 - \phi_2) - \phi_2]/\phi_2^2, \quad (12)$$

where v_e is the effective number of chains connecting crosslink points, and v_e/V_2^0 is the crosslink density of the polymer, which must be established by an independent method. Unfortunately the values of χ determined from swelling measurements are less reliable than those obtained from other observable physical properties [43]. Examination of χ-parameters collected in Tables $3-22$ of Ref. 43, however, show that even those determined by the other methods are not precise enough to permit good correlation of χ with the molecular structure of the liquid sorbed by a given polymer.

The solubility parameter theory (Eq. 4), first proposed by Hildebrand [21], was combined with the Flory-Huggins theory [43] to produce yet another means for determination of χ.

$$\chi = V_1^0(\delta_{pol} - \delta_{liq})/RT - z \, \Delta w_s/k, \quad (13)$$

where the solubility parameters δ_{pol} and δ_{liq} are measures of the intermolecular attractions in the polymer and solvent respectively. As stated earlier the solubility of a polymer in a liquid is taken as maximal when $\delta_{pol} = \delta_{liq}$. Usually the constant entropic contribution, $-z\,\Delta w_s/k$, has values [55] between 0.2 and 0.5.

Values of δ for many liquids have been reported, and these have been recorded in extensive Tables [27, 28, 32, 34, 56–59]. The availability of these data provide an easy means of estimating δ. The solubility parameter theory also has serious shortcomings, however, which limits further the reliability of thermodynamic properties computed by the combination of both theories. Nevertheless it does provide useful qualitative, if not quantitative, descriptions of polymer-solvent systems.

2.3 Forces of Molecular Association that Affect Swelling

Explicit in the Flory-Huggins theory [22–26] is the realization that forces of molecular association are responsible for the finite change in energy that accompanies transformation of pure polymer and pure liquid to a solution of the two components. This change involves rupture of a finite number of molecular liaisons between like molecules in the initial state and formation of a finite number of liaisons between unlike molecules in the end-state. The magnitude of energy change depends on the summation of these forces. There is uncertainty, however, in the definition of the kind and number of liaisons per unit species in the original states, especially that extant in the polymer, which may be crystalline or amorphous but usually is a mixture thereof. The energy required to disrupt the crystalline domains is greater than that required for disruption of amorphous domains. Moreover polymers classified as crystalline contain a finite amount of amorphous region, and conversely polymers classified as completely amorphous contain chain ordering even in solution [60]. Therefore it is difficult to define the initial state or even the true average number of nearest neighbors with respect to a given monomer unit. Also uncertain is the number of liaisons between monomer unit and small molecules in the final state, which presumably must relate to the molecular architectures of these associated species.

Unless the nature and number of the liaisons in the initial and final states are known with certainty, the reliability of the χ-parameter (based on Eq. 7 and relationships derived therefrom) suffers accordingly, even with the most accurate thermodynamic methods for measuring colligative physical properties of polymer-liquid systems. It would be well, therefore, to develop methods for defining the mode of complexation at the initial and final states on a molecular basis. Elucidation of the molecular nature of these complexations at gel-saturation (or in true solution) is an end-objective of the work described in Sect. 3 of this review.

Perhaps the first physical evidence that the force of association between polymer and sorbed liquid is indeed significant came from vapor depression data; for example the vapor pressure of toluene in a solution of 0.01 moles of poly(styrene) (number average molecular weight 290,000) per mole of toluene is 10 fold less than the expected value calculated on the basis of vapor pressure depression in solution [43]. That this force is strong enough to affect the rheological properties of such solutions was reported by Shrag [61] and also by Lodge [62], who studied oscillatory flow birefringence of polymer solutions in high-viscosity solvents. They showed thereby that addition of only small amounts of polymer affect abnormally the solvent viscosity, owing to induced ordering of the surrounding molecules that apparently extends even beyond the first layer of adsorbed molecules. Consistent with this concept of induced ordering are the observations of Bastide [63, 64] and Edwards [65]. Their findings indicate that when high-molecular-weight polymer (MWt > 10^6) is plasticized with low-molecular-weight polymer (MWt < 10^3), and the system is stretched at its rigidification temperature to effect orientation of the high-molecular-weight component, the low-molecular-weight component is similarly oriented, which implies multidentate association of the repeat units in the low-molecular-weight component with the repeat units in the high-molecular-weight component.

The many studies of solvent effects in organic chemistry [66] have shown that liaisons between reactant molecules and the molecules of the solvent affect significantly the rate and even selectivity of reactions that occur in solution. These liaisons usually are between the functional groups in the reactant molecules and the functional groups of the solvent. This is especially apparent in those cases that involve strong associative forces such as hydrogen bonding. The same is true for polymer-liquid systems. Thus, poly(ethylene oxide) is readily soluble in water at room temperature owing to hydrogen bonding of the oxygen atoms in the polymer with the hydrogen atoms in the water. When the temperature of such solutions in raised to about 45 °C, however, the polymer separates from solution, owing to thermally induced dissociation of hydrogen bonding.

Dole [67, 68] observed that small molecules associate firmly with "functional" sites in polymeric materials. He studied sorption of water and organic vapors by synthetic high molecular weight polymers. He noted that the ratio (a/N) of sorbed molecules (a) to available adsorption sites (N, i.e. the total number of functional groups in the polymer) increases with the partial pressure of the vapor being sorbed. In the case of water sorption at 100% RH by poly(N-vinyl 2-pyrrolidinone) a/N was almost one, and in the case of water sorption by poly(vinyl methoxyacetal) [68] it was even greater than one.

Later Yokoyama and Hiraoka [69], who studied evaporation of aqueous solutions of poly(acrylic acid), observed that there is a well-defined ratio of adsorbed water molecules per carboxylic acid group, and that this ratio increases with percent neutralization of the acid groups, such that a/N increases from one at 0% neutralization to four at 100%.

Fowkes [29], who studied the solubility of chlorinated poly(vinyl chloride) in various liquid esters, noted that the solubility of such polymers in these solutions and the intrinsic viscosities thereof decrease with temperature. He interpreted this to mean that the increase in temperature caused dissociation of liaisons formed between acidic hydrogens in the polymer and basic ester groups of the solvent. His own observations and those reported by others, as discussed above, led him to articulate the hypothesis that the "true" solute in polymer-liquid solutions is not the "naked" polymer, but rather it is the polymer "adorned" with solvent molecules that are essentially immobilized by adsorption to the polymer. To be sure these molecules are in exchange equilibrium with the non-adsorbed molecules that comprise the solution. In the above cases the increase in temperature causes displacement of the solvent-polymer-complex equilibrium in favor of dissociation into polymer and liquid, with concomitant self-association of polymer.

Additional evidence that monomer units in the polymer do "complex" with molecules of the solvent comes from studies of the gel-formation that occurs when polymer solutions are cooled well below room temperature [70–72] to induce spinodal phase separation. The polymer does not redissolve until the temperature is raised well above the formation temperature of the system. These gels are supported by interconnected rigid polymer domains, at least part of which are in the form of "helical bands" [72–74]. In the case of isotactic polystyrene these "helical bands" consist of 3_1-helices, which self-aggregate to form crystalline domains. Gelation also occurs with atactic polymer, presumably caused by some form of self-association that does not lead to the well-organized domains that are found for isotatic polymers [74–78] such as isotactic poly(styrene), isotactic poly(acrylates), and also for poly(N-vinylcarbazole), and poly(phenylene oxides). In such cases the structure of the helices that comprise the ordered domains is dependent upon the solvent in which the gel is formed [76–78]. The investigators of this phenomena ascribed the cause of association to dispersion force interactions between the functional groups in the polymer segments and the solvent molecules, such that the distance between segments is influenced by the size and shape of the solvent molecule.

The existence of a solvent-polymer complex in the "helical aggregate" domains was verified by Guenet and co-workers [79–83], who studied the thermal properties, enhanced small-angle X-ray scattering, neutron scattering, and electron photomicrographs of these ordered domains. They showed that a solvated 3_1 helix is formed initially and that this is transformed in turn to a more stable but markedly less solvated 3_1 helix on aging. They concluded that these ordered domains have a structure reminiscent of nematic polymers (i.e. liquid-crystalline material) associated with solvent molecules that affect chain-chain interactions [82]. They showed also that these domains have well-defined ratios of "guest" solvent molecules to monomer units in the "host" polymer. They suggested that the local organization of the solvent molecules around the polymer chain to form a "crystalline" complex strongly governs the ability for physical gelation.

Guenet suggested that this ratio may represent a parameter characteristic of the composition of the helix, which he designated as $\bar{\alpha}$, referring to the number of adsorbed molecules per monomer unit of polymer that separated from solution. If one accepts that the force of association between polymer and solvent is as significant in solution (as suggested by Shrag) as it is in the helix (as suggested by Guenet), then it should be possible to determine the ratio of adsorbed molecules per repeat unit of polymer in the non-crystalline region as well as in the crystalline region of such gels. Comparison of this ratio with the Guenet ratio ($\bar{\alpha}$) would help elucidate how nucleation and propagation of the ordered helical domains come about. To establish such a ratio requires an in-depth study of polymer swelling, which in turn requires a reliable, easy-to-use method for measuring polymer swelling quantitatively.

2.4 Methods of Measuring Polymer Swelling at Liquid Saturation

It is well-known that the volume of a given liquid sorbed at constant temperature by a cross-linked polymer varies inversely with the crosslink density of that polymer [1, 20, 84–90]. It is equally well known [1, 20, 39, 91–94] that determination of polymer swellability was, and still is, a very time-consuming procedure that often yields results of only limited reliability, owing primarily to the ill-defined dimensions and/or inability to measure accurately the weight of the gelled polymer sample at equilibrium swelling in excess liquid. The early investigators of poly(Sty-*co*-DVB) swelling [1, 39] waited 3 to 7 days for copolymer samples to equilibrate with solvent before removing the sample for damp-drying between absorbent materials and subsequent measurement of liquid uptake, either gravimetrically or volumetrically. The reproducibility of the measured volume or weight of the damp-dried gelled polymer samples was poor at best, and consequently as many as 40 replicated determinations were averaged together to obtain one reliable value [39] for the swelling ratio $(V_g - V_p)/V_p$, where V_g and V_p are the volumes of the polymer in the gelled and dry states, respectively.

Volumetric methods based on liquid displacement have been developed by Bobin [95], Garvey [96], Zhuravlev et al. [97], Schreiber et al. [98], and Buckley [84], which generally are more precise for measuring swelling than the earlier methods based on dimensional changes. Chicklis and Grasshoff [99] developed an optical device for measuring liquid uptake, which appears to improve the sensitivity of earlier devices for measuring swelling of very thin (1 to 10 μm) films, and the reduced dimension lessens the time required to attain equilibrium swelling before final measurement.

More recently it has been shown that it is possible to measure the dimensional changes of microbeads, upon relatively rapid saturation in a test liquid, by means of photomicrographic techniques [100–102]. Although considerably more precise than those noted above, these measurements are still very time consuming and require expensive instrumentation. Methods for measuring swelling volumetrically in "batch" quantities of microbeads have also been reported [103, 104], but these too entail a long and time-consuming protocol for each determination.

Owing to higher surface-to-volume ratio, the time to attain gel-saturation using microbeads or powders is even less than that for very thin films. Polymers in these forms have been used in gravimetric methods, which are especially useful for measurement of sorption of gases and vapors [7, 45, 105–116]. Peppas [117] has shown, however, that this may not be such an advantage in the case of liquids, because of a so-called "relaxation effect" [93], which results in an "overshoot"; i.e. the volume of sorbed liquid becomes maximal in a relatively short time as expected, but then decreases slowly thereafter to an asymptotic limit that is reached in about a month. Peppas attributes this slow decay in sorption capacity to "relaxation" of polymer chains. Such an "overshoot" was also observed much earlier by Boyer [39], who used relatively larger polymer samples, but the magnitude of the effect observed by Peppas was greater, owing to the much shorter time to reach maximal saturation, which minimizes slow "relaxation" during the interim. The total time required to attain the stable end state, however, was similar.

This "overshoot" or "relaxation effect" may instead be an artifact resulting from slow extraction of soluble polymer or residual trapped sorbates. If this were true, the "overshoot" could be eliminated through preconditioning of the sample by prolonged extraction in one or more good solvents before measuring the sorption capacity in a test-liquid. This time-consuming modification added to already very time-consuming procedures makes the combined protocol almost prohibitive except to a very patient (and long-lived) investigator. Nevertheless a preconditioning procedure that removes all effects of prior history of polymer formation, subsequent exposure to vapors, and any residual liquids, is mandatory regardless of the method of measuring swelling. From the standpoint of available time, however, one can not afford to make month-long investments in the preconditioning of a given sorbent polymer unless that material can be re-used many times thereafter without the need for further repetitions of the elaborate conditioning.

It must be concluded, therefore, that none of the above available methods is suited to serve as a standard protocol for an analytical study of polymer swelling. Fortunately, a convenient, easy to use, very reliable method and reproducible for measuring polymer swelling has been developed in the 3M laboratories as an offshoot of a product-oriented research effort. This method is discussed in the following section.

3 Sorption Studies Using Particulate Polymer Enmeshed in PTFE

3.1 Development of the Analytical Method

My interest in undertaking meaningful studies of polymer swelling was awakened when I discovered that it was possible to make microporous composite membranes from dough-like mixtures of particulate matter and aqueous polytetrafluoro-ethylene (PTFE) emulsions via a work-intensive kneading process, preferably by use of an ordinary rubber mill [118–121]. In so doing the particles become separated and enmeshed individually in an entangled network of PTFE microfibers, as indicated by the example shown in Fig. 1, which is a set of four SEM photomicrographs of Sephadex (i.e. crosslinked dextran) particles enmeshed in PTFE. With the obvious exception of those particles located in the plane of the freeze-fracturated edge (Figs. 1B and 1C), the particles are trapped permanently in the entangled network of PTFE fibers; particles in the plane of fracture are eliminated simply by agitation in a suitable fluid.

These microporous composites are tough, but soft and very conformable, leather-like films, the chemical and physical properties of which are determined primarily by the choice of the major component (i.e. the enmeshed particle, usually more than 85% by weight) [121–124]. The minor component (PTFE), however, confers physical integrity upon the film without affecting adversely the chemical properties of the major component. Since the enmeshed particles are distributed

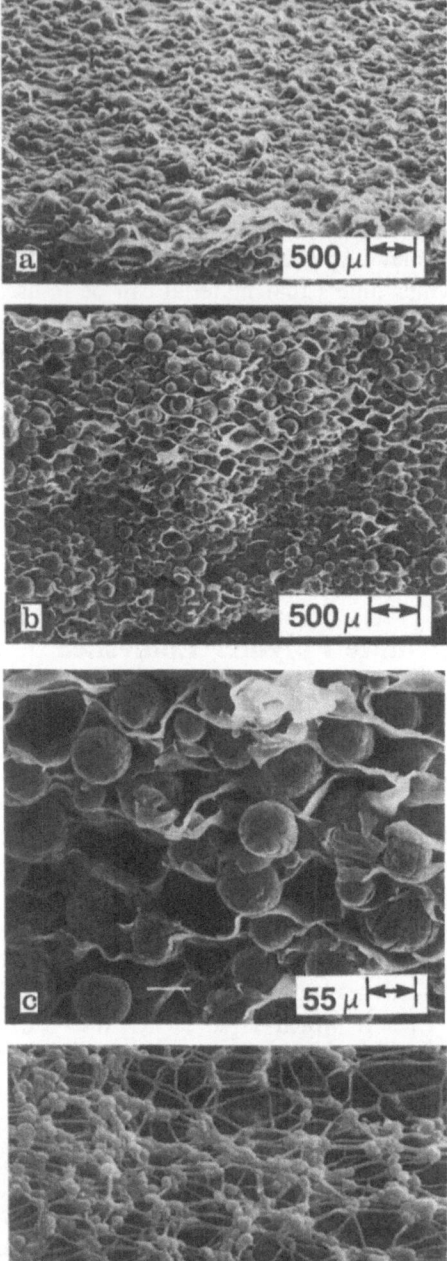

Fig. 1a–d. SEM photomicrographs of a microporous composite sheet consisting of Sephadex (crosslinked dextran) microspheres (approx. 80% by weight) enmeshed in PTFE fibers. (**a**) Top surface viewed at a 45° angle from the horizontal plane. (**b**) Freeze fractured edge, viewed at a 0° angle from the horizontal plane. (**c**) Tenfold enlargement of "B". (**d**) Forty-four-fold enlargement of "C" showing an area of PTFE entangled microfibers and residual non-fibrillated PTFE microspheres

uniformly throughout the bulk of the composite structure as noted in Fig. 1, and they are all uniformly accessible to permeating fluids *without channeling*, these microporous composition are ideally suited for uses such as (1) chromatographic devices (alumina, silica or titania particles) [122, 123], (2) catalytic membranes (Raney nickel or palladium-on-charcoal powders) [123, 124], (3) reactive membranes for selective removal of either solvated ions (zeolite or ion-exchange-polymer particles) [123], or organic molecules (activated charcoal or heterocyclic chars) [123, 124], and (5) sorption of liquids, such as water (Sephadex, polyacrylamide, or cellulosic particles) [123], silicone oils (microfine silica particles) [123], or organic liquids (cross-linked polystyrene or polyacrylate particles) [123, 125].

One of many potential uses for such compositions is a bio-effective membrane designed to replace pigskin in burn therapy [120, 121]. The active component in this product is a water-swellable particulate material, such as crosslinked dextrans or polyacrylamides, that serves as a sterile barrier to microorganisms and as a poultice that absorb toxins from the wound area. In the course of developing such a product, which required careful extraction to remove potentially harmful ingredients, I noticed that the volume sorbed per gram of membrane at saturation in excess water at 23 °C is quite reproducible [19] (Figs. 2 and 3), despite that the kinetics of water sorption is history-dependent (i.e. dependent upon factors that affect wetting properties [126–128]). Moreover the original weight of the composite film (usually >1 g) is restored to within ±0.1 mg after the water-saturated film is dried in an evacuated oven kept at 100 °C. These results were very encouraging from the standpoint of polymer swelling studies, because they indicated that the use of these films might be ideal for studying swelling of the

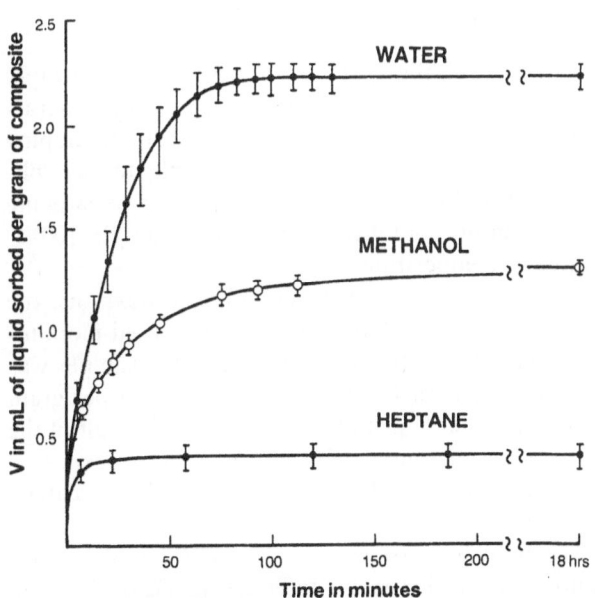

Fig. 2. Kinetics of liquid sorption in 10 replications using samples cut from the same microporous composite sheet. The sorbent particles were Sephadex microspheres (Example shown in Fig. 1)

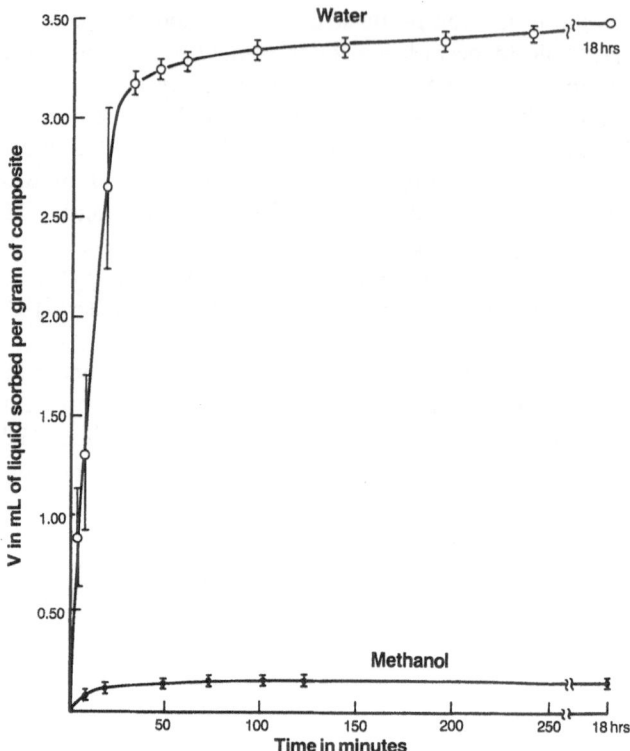

Fig. 3. Kinetics of liquid sorption in 10 replications using samples cut from the same microporous sheet. The sorbent particles were comminuted crosslinked polyacrylamide (approx. 80% by weight) enmeshed in PTFE microfibers

enmeshed particles. It appeared to offer the best of two worlds, namely rapid equilibration with liquid (characteristic of powders) and ease of handling (characteristic of a robust polymeric film). Moreover, it also offered the hope of developing a method for measuring polymer swelling based on a reusable test-sample, provided that the composite sample could be put through many cycles of swelling to saturation, cleaning in a suitable liquid, and then drying at, for example, 100 °C without loss in weight or change in composition.

In the case of biodegradable particulates such as cross-linked dextrans, the results were less than satisfactory [19]. The sorption capacity decreased monotonically with the number of swelling cycles (i.e. total time that the sample was allowed to remain in contact with water) such that the volume sorbed per gram of membrane after fifteen days in non-sterile distilled water was about half of the original sorption capacity. SEM photomicrographs before and after this exposure showed that large chunks of each originally spherical particle had been "eaten" away, presumably by biodegration.

The sorption capacities of carbohydrate particles for various liquids is also affected by swelling in methanol [19], which appears to cause significant reaction with functional groups. Thus, the sorption capacity of the dextran-PTFE sample before swelling in methanol (Fig. 2) was 2.23 ± 0.06 mL/g for water, whereas it was only 0.40 ± 0.04 mL/g for heptane (indicative of wetting to fill the interstices of the composite membrane but not swelling of the enmeshed particles therein). After the membrane was cycled through swelling in methanol and then retested in water and in heptane, the average sorption capacity for water decreased to 1.57 ± 0.05, whereas that for heptane increased to 1.30 ± 0.04 (indicative of significant particle swelling).

On the other hand membranes made with polyacrylamide particles did not exhibit the above chemical sensitivity, despite that the kinetics of water uptake was markedly increased by prewetting in methanol [19]. These membranes could be made to undergo many swelling and drying cycles without measurable change in the sorptive capacity for either solvent. The above results stress the importance of the need to avoid test-liquids that cause permanent changes in the enmeshed particles, which render the results no longer characteristic of the original composition but instead that of the chemically-modified product.

Although no evidence was noted in the above studies that enmeshed particles are lost when such composite membrane are allowed to swell only 2 to 3 fold, one may well ask if the same is true for membranes that swell 20 to 30 fold. Conceivably this magnitude of swelling could destroy the physical integrity of the PTFE network. To test this possibility I chose a composite membrane consisting of Super-Slurper (cellulose chain-extended by graft polymerization of acrylonitrile, subsequently hydrolysed with aqueous base) enmeshed in PTFE (15% by weight). Two membrane samples of essentially the same size and shape were allowed to swell to saturation in deionized water (sorption capacity over 90 mL/g), dried to its original weight and then reswelled to saturation [19]. The cycle was repeated a total of five times, and it was observed that the saturation limit decreased monotonically owing to ion-exchange with fresh deionized water. The sample was then acidified with HCl to ensure complete conversion to free carboxylic acid at which state the absorption capacity became only 1.7 mL/g of membrane. Reneutralization with base caused the sorption capacity to resume its original level over 90 mL/g in distilled water. The sample was then dried to constant weight, which was found within 0.05% of its original weight. These results proved that few if any of the enmeshed particles were lost even when exposed to extreme swelling and drying cycles.

Why the enmeshed microparticles are not lost despite the more than 90 fold swelling to saturation in deionized water becomes clear when one examines in Fig. 4 the fiber structure in the liquid-saturated and air-dried samples [19]. Figure 5 is a set of four SEM photomicrographs of the fiber structure of the liquid-swollen sample after it was freeze-dried. The views of the top surface (A) and freeze-fractured edge (B) show the extent of expansion of the fibrous cells, each of which had contained a polymer particle. View (B) shows that virtually all of those particles that had occupied the cells in the fracture plane were lost despite gentle handling of the fractured film. Figure 6 is a similar set of four SEM photomicrographs of the sample that had been cut to the same size and shape in its water swollen state and then allowed to evaporate to dryness in air. The views of the top surface

Fig. 4. Photograph of a sample that was cut from a water saturated composite sheet consisting of Super-Slurper (cellulose chain extended with hydrolysed polyacrylodnitrile) particles (85% by weight) enmeshed in PTFE microfibers. The smaller sample, which had been cut to the same size and shape from the same water saturated composite film, was allowed to evaporate to dryness at ambient room conditions. The original size and shape (represented by the larger sample) was re-established when the smaller sample was allowed to re-swell to saturation in very dilute aqueous NaOH

(A) and fractured edge (B) show the dense intimate packing of particles and associated fibers. The enlarged views (C and D) of (B) indicate that the extended fibers, which enmesh the soft gelled particles at liquid-saturation, adhere to these particles as they decrease in volume with progressive evaporation. It is this property of PTFE microfibers that prevents loss of enmeshed particles, despite the enormous physical changes that occur during the swelling/drying cycles.

The final potential uncertainty concerns "overshoot" or "polymer relaxation" observed by Boyer [39] and by Peppas [93] (see above). We verified that such "overshoots" are observed, when freshly made composite membranes consisting of poly(styrene-*co*-divinylbenzene) [hereafter referred to as poly(Sty-*co*-DVB) or $(Sty)_{1-x}(DVB)_x$)] particles enmeshed in PTFE microfibers (Fig. 7) are made to swell in a liquid such as toluene or chloroform. We noted, however, that as much as 5 to 10% of the original weight of enmeshed particles is extracted during this very prolonged initial swelling cycle, but little if any weight is lost in subsequent swelling/drying cycles [125]. When these "conditioned" films are made to reswell in the same liquid, an "overshoot" does not occur. An "overshoot" is observed, however, when such a "conditioned" film is made to reswell in a second liquid that has a much weaker affinity for the polymer than the preceding liquid. This latter "overshoot" is in fact caused by a trace amount of residual first liquid that was not removed during the drying cycle even after 24 hours in vacuum at 100 °C. Such "memory" effects are avoidable, however, by "cleaning" the sample in a poorer solvent such as acertone before re-swelling in the next test-liquid. Extraction (preferably for several hours) in excess acetone removes the previous adsorbed liquid by molecular displacement, and the acetone is eliminated quantitatively thereafter in vacuum at 100 °C within one hour.

When the composite membranes have been preconditioned and "cleaned" as described above, the kinetics of sorption as well as the sorption capacity at the asymptotic limit is highly reproducible [125] as shown in Fig. 8, which records as

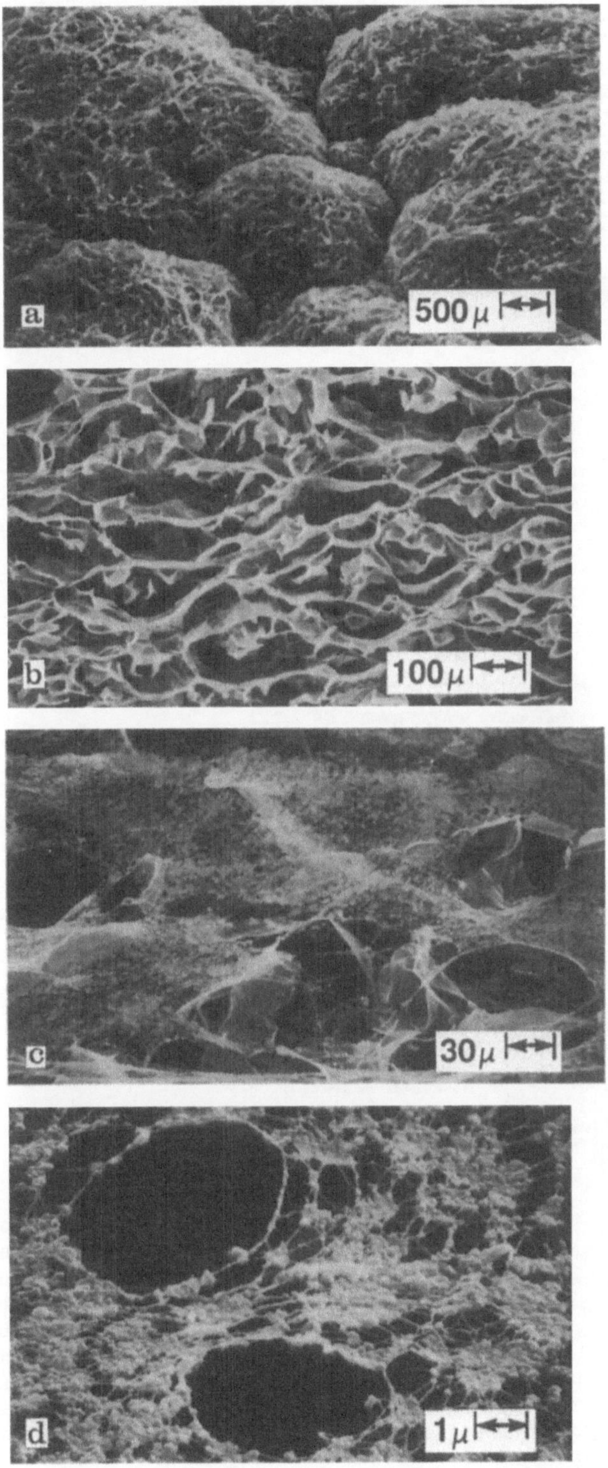

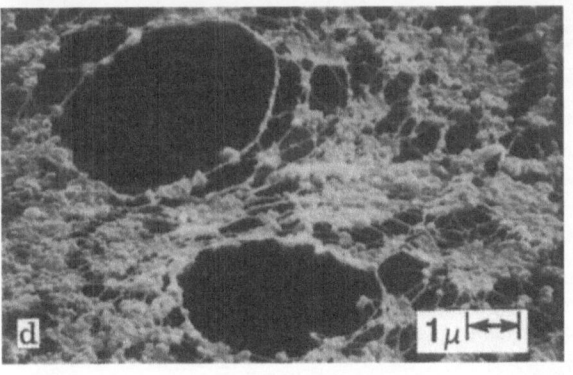

Fig. 5a–d. SEM photomicrographs of the water-saturated sample (larger sample in Fig. 4) after it was freeze-dried. (a) Top surface viewed at 45° from the horizontal plain. (b) Freeze-fractured edge viewed at 0° from the horizontal plain. (c) Three-fold magnification of B. (d) Thirty-fold magnification of C, showing the magnitude of expansion of the PTFE microfibers caused by the >90 fold swelling to saturation in water

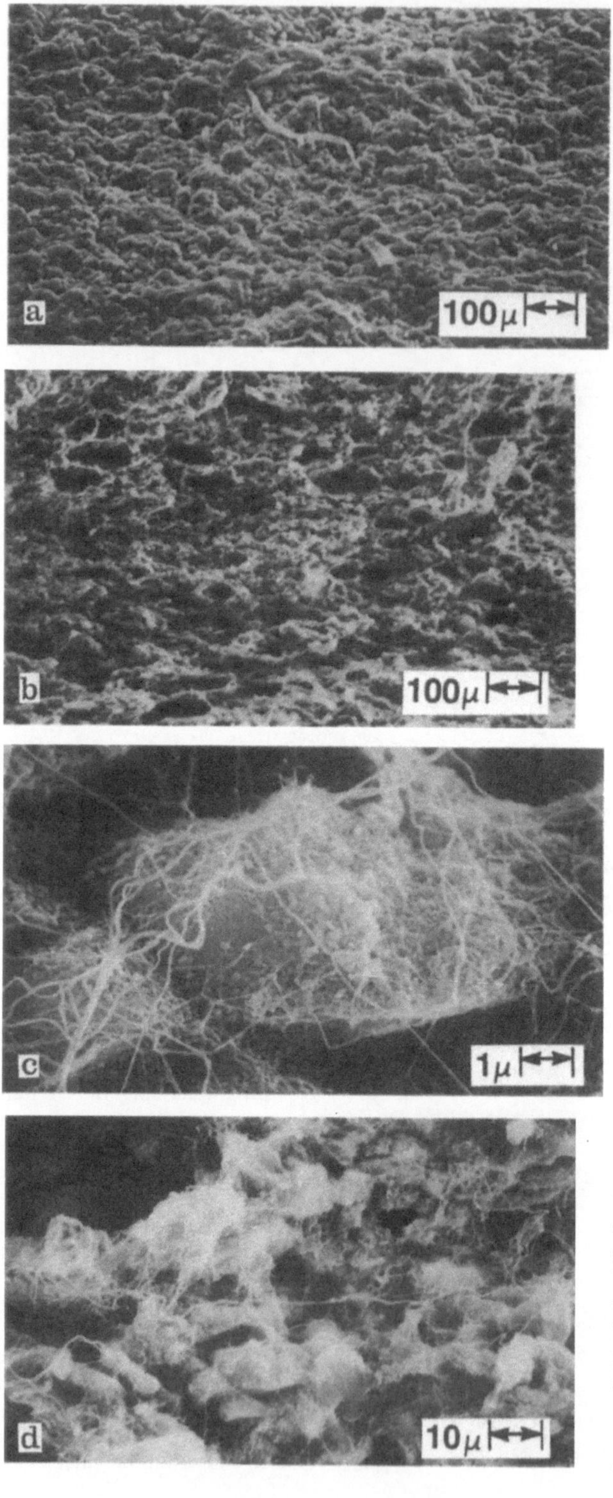

Fig. 6a–d. SEM photomicrographs of the air-dried sample (smaller sample in Figure 4). (**a**) Top surface viewed at 45° from the horizontal plane. (**b**) Freeze fractured edge viewed at 0° from the horizontal plain. (**d**) Ten fold enlargement of B. (**c**) Tenfold enlargement of D

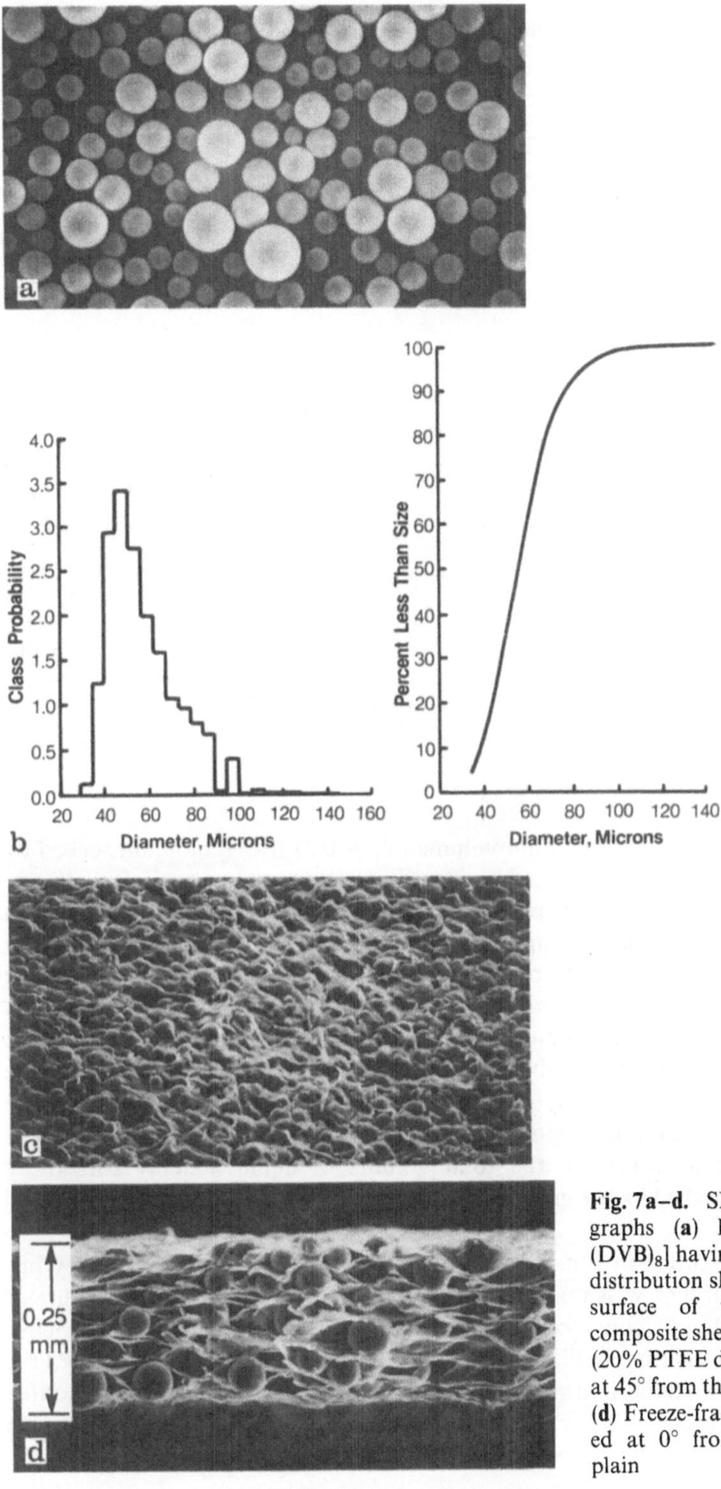

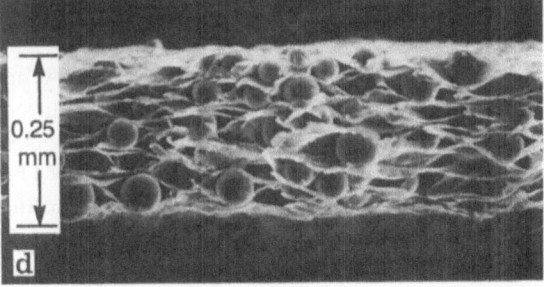

Fig. 7a–d. SEM photomicrographs (**a**) Biobeads [(Sty)$_{92}$ (DVB)$_8$] having the particle size distribution shown in b. (**c**) Top surface of the microporous composite sheet made therefrom (20% PTFE dry weight), viewed at 45° from the horizontal plain. (**d**) Freeze-fractured edge, viewed at 0° from the horizontal plain

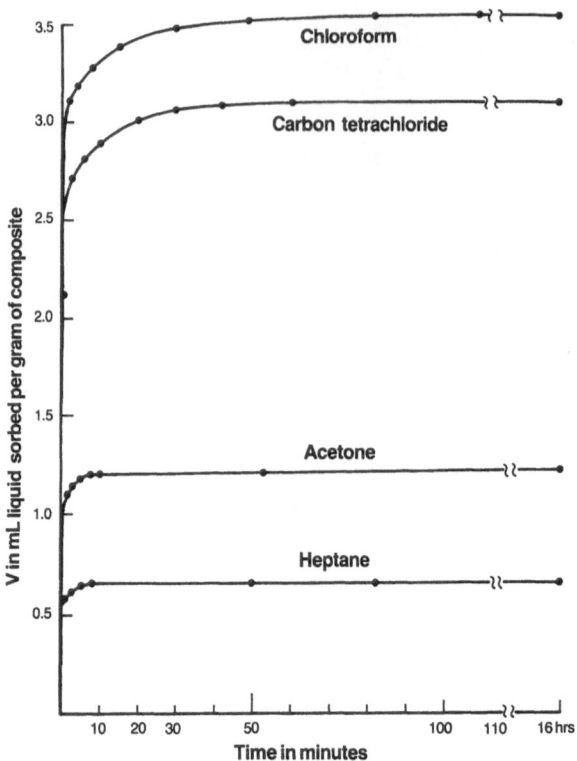

Fig. 8. Kinetics of liquid sorption in 10 replications using samples cut from the same composite sheet consisting of poly(Sty-*co*-DVB) microspheres (80% by weight) enmeshed in PTFE microfibers (similar size-distribution to that shown in Fig. 7)

a function of time the average total volume ($V_t \pm 0.01$ mL/g) of liquid sorbed by ten one-gram swatches cut from the same microporous composite membrane consisting of poly(Sty-*co*-DVB) particles (80% by weight) enmeshed in PTFE. In the case of heptane, the asymptotic limit is attained within 10 minutes at $V_s = 0.69 \pm 0.01$ mL/g. Since this liquid wets but does not swell significantly crosslinked poly(styrene), 0.69 mL/g represents the volume of liquid required to fill by capillary action the interstices of one gram of *this* composite film (i.e. the void volume characteristic of the film and all swatches cut therefrom). In the case of liquids that swell poly(styrene) significantly, however, the asymptotic limit, V_s, is attained much more slowly, but usually within two hours. At this state of liquid-saturation the total volume of liquid actually sorbed by the particles enmeshed in one gram of this composite film is given by the difference ($V_s - 0.69$).

The results recorded in Fig. 8 not only demonstrate good reproducibility but they show as well that V_s is markedly dependent upon the molecular structure of the test-liquid. I was convinced by these results that this simple gravimetric procedure for measuring swelling of polymer particles enmeshed in inert PTFE fibers represents a reliable and very convenient method for collecting large amounts of swelling data rather quickly, and that it would now be practical to undertake systematic swelling studies aimed at correlating swelling power with the molecular structures of the sorbed liquid and sorbent polymer. A good first step in this direction would be the establishment of a linear relationship that correlates sorption capacity, S, per gram of polymer at liquid-saturation with the known crosslink density of a sorbent polymer.

3.2 Correlation of Polymer Swelling with Crosslink Density

3.2.1 Theoretical Considerations

Flory pointed out [20] that a close analogy exists between swelling equilibrium and osmotic equilibrium. In the case of a crosslinked polymer at thermodynamic equilibrium with excess liquid, the gelled network structure performs the multiple role of solute, osmotic membrane, and pressure-generating device. Each segment of length "l" between crosslink junctions is completely associated with solvent molecules; thus the segments are forced to form elongated configurations in a random mode to occupy the solvent volume. Flory suggested that a force akin to "elastic" retraction in rubber develops, which is in opposition to the swelling process. This force increases with swelling, while the dilution force decreases, yielding to a balance of chemical potentials of the solvent inside and outside the gelled polymer. Ultimately, the restraining force of gelled network becomes equal to the osmotic pressure π, and a state of equilibrium at V_s is established.

In view of the observation that the kinetics for polymer swelling by a liquid is Case II rather than simple Fickian (see above or Ref. 10–13), it seemed to me that initially the force in opposition to the swelling process is more like a string-force than an elastic spring-force, i.e. once the force of self-association that holds the polymer together is overcome by adsorption of small molecules, which imparts local mobility to the segment, swelling around that segment occurs rapidly without significant increase in resistance until the segment approaches the end of its "tether", i.e. after all of the monomer units in the polymer segments have been saturated with adsorbed molecules. Thereafter the force of resistance to further swelling is akin to that of elastic retraction as the system swells to the turgid end-state, representing equilibrium between the two opposing forces. Thus the greater the number of adsorbed molecules per accessible monomer unit of polymer, the more extended will be the polymer segment between crosslinked junctions, and the greater will be the sorption capacity per gram of polymer.

If one accepts the above small modification of the Flory model for equilibrium swelling, and also assumes that the turgid end-state is described by the van't Hoff relationship:

$$\pi V_s = nRT,$$

where n is the number of moles of polymer segments in one gram of polymer, T is the absolute temperature, R is the universal gas constant in appropriate units, and V_s is assumed to be a linear function of "l", the average length of the segments between crosslink junctions, which in turn is proportional to the average number

(λ) of back-bone carbon atoms in the polystyrene segment of average length "l", then it is possible to proceed toward a relationship that gives the volume (V_s) sorbed by a monomer unit of polymer at liquid-saturation as a function of the macrostructural cross-link density of that polymer. Since π is inversely proportional to the radius (if V_s is considered to be in the form of a sphere) or to an edge (if V_s is considered to be in the form of a cube), it follows that π, the force per unit area in a given direction [that restrains the gel volume to the equilibrium volume $V_s = f(\lambda)$] must be inversely proportional the the the cube root of V_s; i.e. π is a function of $\lambda^{-1/3}$ [hereafter written as $\pi = f(\lambda^{-1/3})$]. Substituting this function for π in the van't Hoff relationship gives:

$$V_s = (nRT)/f(\lambda^{-1/3}), \quad \text{or} \quad af'(\lambda^{1/3}) + b,$$

where a and b are constants (at constant temperature). Substituting V_s into the swelling ratio, $S = (v_s - V_0)/V_0$, gives:

$$S = (1/V_0)\,[(af'(\lambda^{1/3}) + b) - V_0],$$

where V_0 is the volume of one gram of dry polymer. Since a, b and V_0 are constants this relationship simplifies to:

$$S = C\lambda^{1/3} + C',$$

where the value of C' is $-C\lambda_0^{1/3}$ and λ_0 is the average number of backbone carbons between crosslink junctions, below which S is zero. Thus, the magnitude of S should vary linearly with $\lambda^{1/3}$ as expressed by:

$$S = C(\lambda^{1/3} - \lambda_0^{1/3}), \tag{14}$$

where the difference $(\lambda^{1/3} - \lambda_0^{1/3})$ is a dimensionless number, Λ, that reflects the relative "looseness" of the polymeric macrostructure, and C is the relative swelling power of the sorbed liquid in ml per gram of polymer.

The relative swelling power, C, of the sorbed liquid is the product of two factors, namely the number, α, of *ad*sorbed molecules per *accessible* monomer unit of polymer at liquid-saturation, and β, i.e. $(M/d_p)/(M_p d)$ where M and M_p are the formula weights of the sorbed liquid and monomer unit of polymer respectively and d and d_p are the respective densities of the liquid and polymer. From the standpoint of interpretation of polymer swelling in terms of molecular structure, α is more meaningful than C, from which α can be calculated by means of Eq. 15.

$$\alpha = M_p Cd/Md_p \quad (\text{i.e. } \alpha = C/\beta). \tag{15}$$

Since the *ad*sorbed molecules are in exchange equilibrium with the rest of the sorbed molecules, α is in effect the average dynamic packing density of the molecules immobilized by adsorption to the *accessible* monomer units of crosslinked polymer at liquid saturation. If the relationship expressed by Eq. 14 is indeed correct as postulated, then α should be independent of the macrostructure of the polymer as measured by λ, which means that α should in fact be a measure of the number of adsorbed molecules per monomer unit of polymer in true solution. The magnitude of α, therefore, should reflect how well the molecular structure of the adsorbed molecules is accommodated by the molecular structure of the monomer unit of polymer, i.e. it should vary directly with the (electronic) affinity of the sorbed molecule for the repeat unit of the polymer and it should be related inversely to the "bulkiness" of the sorbed molecule because of steric hindrance (from the standpoint of volume exclusion with respect to the finite available solid angle "space" at the adsorption site).

3.2.2 Experimental Verification of Theory

The gravimetric method for measuring liquid uptake by polymer particles enmeshed in PTFE microfibers [125] (see Sect. 3.1), afforded us a simple, very convenient, but very reproducible analytical method for testing the validity of Eq. 14. To this end, six microporous composite films that contained $(Sty)_{1-x}(DVB)_x$ particles, each with a different level of x, were made according to the procedure already described [19]. The particles were obtained from Bio-Rad Laboratories, who determined the mole fraction, x, of DVB in each lot of polymer on the basis of the ratio of ethylbenzene to diethylbenzene isolated via thermal degradation of representative samples from these polymer lots [129]. After the films were "conditioned" by exhaustive extraction in toluene, chloroform and finally in acetone, a swatch of distinctive shape (to permit easy identification of x for the particles contained therein) and weighing about one gram was cut from each film. This set of six swatches was then allowed to swell to saturation in the test-liquid; the weight of liquid sorbed thereby was determined gravimetrically by difference and with suitable correction for the volume of liquid in the interstices unique to each film sample; thereafter the swatches were "cleaned" by extraction in acetone and then dried to constant weight at 100 °C in a vacuum oven; this weight was within ± 0.1 mg of the original weight, which indicated that the swatch was now ready for swelling in the next test-liquid. The volume of liquid, S, sorbed per gram of enmeshed particles in each of the six samples so determined was then plotted versus the corresponding crosslink density of the particles contained in each sorbent swatch.

The average number (λ) of backbone carbon atoms in the polymer segments between crosslink junctions is by definition the total number of backbone carbon atoms divided by the total number of these atoms that form a crosslink junction, i.e. bonded covalently with three or more polymer segments. In the case of $(Sty)_{1-x}(DVB)_x$, λ is given by $(1 + x + a)/x$, where "a" is the contribution of the phenyl group in DVB to the average length of the segment. Because this group serves as a knot that puts a kink in the chain and x is small, "a" can be ignored as a first approximation so that $\lambda = (1 + x)/x$. The crosslink density, therefore, is the inverse of λ, i.e. $x/(1 + x)$, and when x is <0.1, λ is equal approximately to x.

Benzene was chosen to be the first test-liquid because the data obtained thereby could be compared directly with the sorption data reported by Staudinger [1].

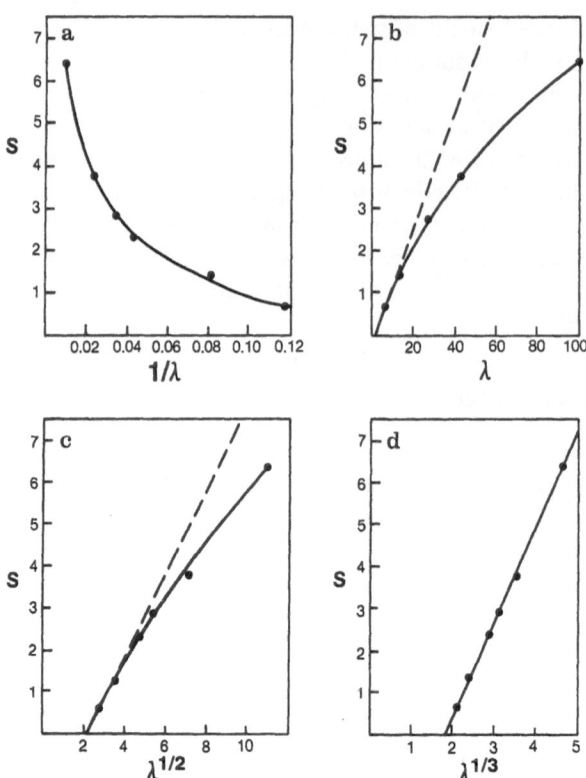

Fig. 9a–d. Correlations of volume (S) of sorbed benzene per gram of $(Sty)_{1-x}$ $(DVB)_x$ particles (80% by weight) enmeshed in PTFE microfibers with (**a**) the cross-link density, $1/\lambda$ (where λ, equal to $1/x$ when x is <0.1, is the number of backbone carbon atoms between crosslink junctions), (**b**) the reciprocal of crosslink density (λ), (**c**) the square root of λ, and (**d**) the cube root of λ.

The sorption data obtained in our laboratory [121] are correlated in Fig. 9 with the cross-link density ($1/\lambda$; Fig. 9A), with the reciprocal of cross-link density (λ, Fig. 9B), with $\lambda^{1/2}$ (Fig. 9C), and with $\lambda^{1/3}$ (Fig. 9D). As expected on the basis of Staudinger's results, the sorption capacity, S in mL/g, decreased monotonically (Fig. 9A) but not linearly with crosslink density, and S increased monotonically (Fig. 9B) but not linearly with the reciprocal of crosslink density (λ), which verifies Staudinger's observation that S does not vary linearly with either the monomer charge ratio (R = DVB/Sty) or the reciprocal thereof. A linear relationship is exhibited, however, when S is correlated with $\lambda^{1/3}$ (Fig. 9B). This result is in agreement with expectation on the basis of Eq. 14, and the assumption that the ratio of DVB to Sty incorporated in the polymer, i.e. R′, is proportional to the corresponding monomer charge ratio (R) made to undergo polymerization by free-radical propagation.

The reproducibility of the swelling results obtained with the above analytical protocol is illustrated in Fig. 10, which is a plot of the data obtained in five such sets of six determinations using toluene as the test-liquid [125]. The average value of S for a given sorbent sample is represented by the center of the empty circles (Fig. 10), and the corresponding range of the data set is indicated by the vertical bar through the respective empty circle. The best line through each set of six data-points obtained in a given experiment were straight lines that were virtually

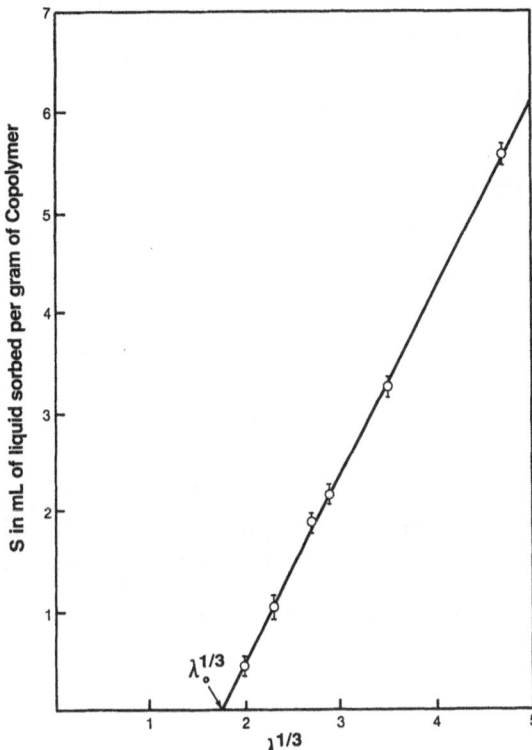

Fig. 10. Correlation of volume S (mL) of toluene sorbed per gram of $(Sty)_{1-x}(DVB)_x$ particles (enmeshed in PTFE microfibers) at liquid saturation with the corresponding $\lambda^{1/3}$ in the polymer sample. The *center* of the *empty circles* represents the average of 5 replicate determinations for a given sample with crosslink density $1/\lambda$. The length of the *vertical bar* through each empty circle represents the range of the data for that set of 5 determinations

superimposable upon the line shown in Fig. 10. Thus the slope $C = S/\Lambda$ is in effect an average value of six determinations of C. (The data recorded in Figs. 9 and 10 represent the results of the first 6 sets of 6 determinations to test linearity with respect to $\lambda^{1/3}$. Since then more than 2,000 such determinations have been made and the techniques for so doing have improved accordingly, with a commensurate improvement in precision).

The results observed with a representative set of test-liquids are recorded in Fig. 11. The square of the correlation coefficient (r^2), determined by linear regression analysis, was in every case > 0.99 when the slope was > 1, but r^2 for those with slopes less than 1 were sometimes less than 0.98, owing to decrease in accuracy as a result of decrease in the difference $(S - v)$ where v is the porosity characteristic of the respective composite film samples. In such cases the determination was repeated to obtain sets of averaged values, which typically exhibited $r^2 > 0.99$.

Consistent with Eq. 14, the data recorded in Figs. 9 D, 10 and 11 show that the straight-line relationships for S as a function of $\lambda^{1/3}$ intersect the abscissa at $\lambda^{1/3} = 1.8 \pm 0.1$. This common point of intersection, $\lambda_0^{1/3}$, appears to be characteristic of the polymer and reflects the crosslink density, $1/\lambda_0$ above which swelling (S) is not measurable by this method of analysis, i.e. when DVB/Sty is greater than about 1/5.

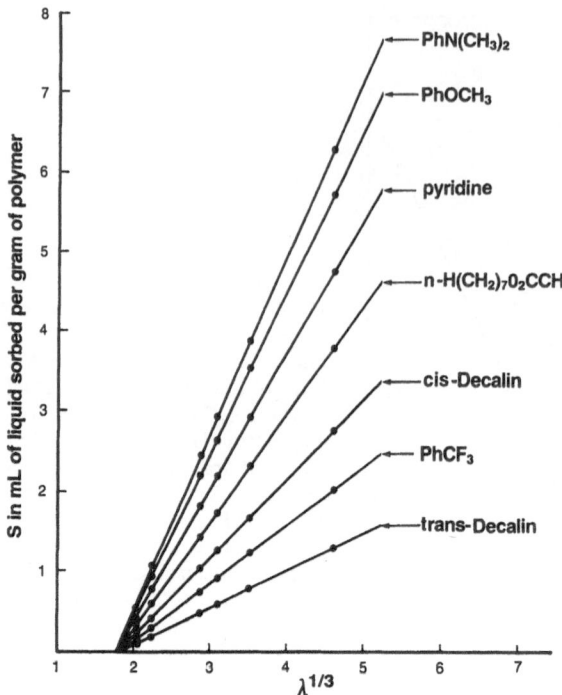

Fig. 11. Typical examples of sorption-$\lambda^{1/3}$ correlations for seven liquids, showing how the slope of these linear relationships are affected by the molecular structure of the liquid

Having established the validity of Eq. 14, this relationship can now be used, in rearranged form (Eq. 16), to calculate the crosslink density $(1/\lambda)$ of poly(Sty-co-DVB) polymers, the swelling ratio of which was already reported or otherwise determined by any other method, provided that C and λ_0 for the sorbed liquid are known.

$$1/\lambda = [C/(S + C\lambda_0^{1/3})]^3 . \tag{16}$$

3.2.3 General Applicability to Poly(Sty-co-DVB)

In order to test whether or not Eq. 14 is general for all poly(Sty-co-DVB) samples made via free-radical polymerization, or unique for the samples obtained from Bio-Rad Laboratories, the data reported about 50 years ago by Staudinger [1] and by Boyer [29] were used to test conformance with prediction [130] on the basis of Eq. 14. In these classical studies of polymer swelling, the increase in volume owing to sorbed liquid was reported by Staudinger as $(V_s/V_0 - 1)$, and by Boyer as V_s/V_0, where V_0 and V_s are respectively the volumes of dry polymer and of liquid-saturated polymer. Since the molar charge ratio R (i.e. DVB/Sty) was less than 0.001 in all cases, R is virtually the same as the DVB mole fraction, y, in the monomer solution made to undergo copolymerization to give the polymer product with composition $(Sty)_{1-x}(DVB)_x$, where x, the mole fraction of DVB in the polymer, is assumed to be a function of y. The magnitude of x, however, was never

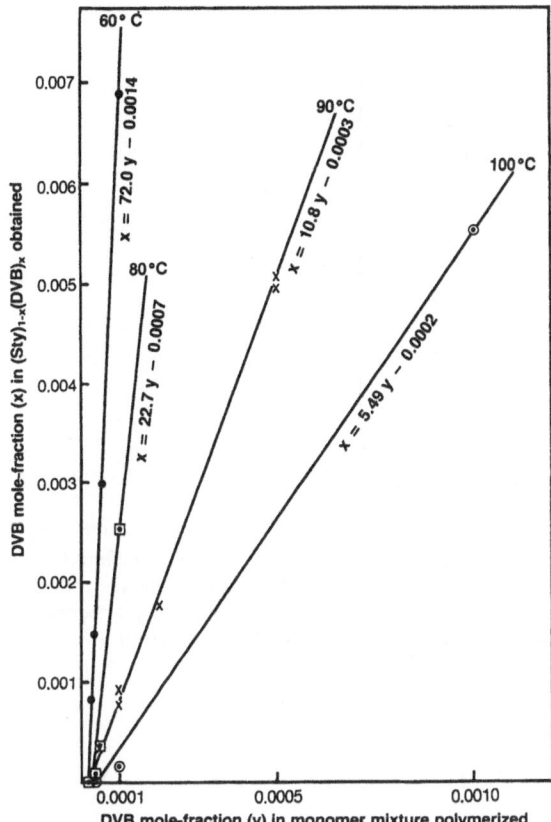

Fig. 12. Correlation of the calculated DVB mole-fraction (x) in the $(Sty)_{1-x}(DVB)_x$ (based on the magnitude of swelling reported in 1935 by Staudinger) with the corresponding DVB/(Sty + DVB) ratio (y) made to undergo polymerization at the temperatures indicated

established by Staudinger simply because the means for doing so was not available at that time. If Eq. 14 has general validity, however, then it should be possible to use their data and the relative swelling power determined by us for the liquids used in their studies, to calculate x (using Eq. 16) for the $(Sty)_{1-x}(DVB)_x$ actually produced by the earlier investigators. The calculated value of x could then be correlated with y to see whether or not the linearity observed over the range of x = 0.01 to 0.11 for Bio-Rad polymers prepared recently also holds true for the polymers prepared 50 years ago, the y of which ranged from 0.0001 to 0.001. If it does, then y should show a linear relationship with x.

The results obtained for this correlation [130] are recorded in Fig. 12, which shows that at each given temperature of polymerization reported by Staudinger, x is a linear function of y as expressed by:

$$x = Ay - B,\qquad(17)$$

where the constants A and B vary inversely with the temperature of polymerization. When A is correlated with 1/T (Fig. 13), an Arrhenius-like relationship is obtained

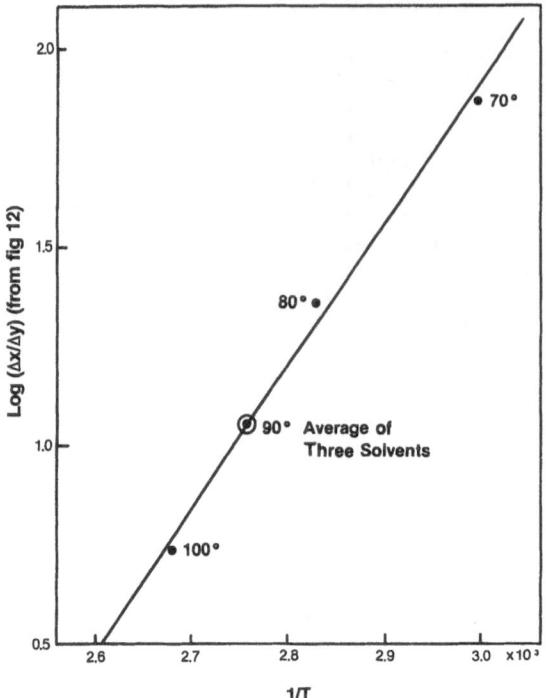

Fig. 13. Correlation of the slopes of the linear relationships shown in Fig. 12 with the reciprocal of the corresponding temperature (K) of polymerization

[130], which is given by:

$$\log A = [(E_2 - E_1)/2.3R]\,(1/T) + K\,, \tag{18}$$

where E_1 and E_2 are the effective "activation energies" for copolymerization of DVB and styrene respectively under the experimental conditions used by Staudinger. From the slope of this line (Fig. 13), the difference $(E_2 - E_1)$ was calculated to be about 8 kcal/mole.

Similar treatment of the swelling data reported by Boyer [39] for his copolymers prepared at 105 °C also show a linear relationship [130], but the values of the A (9.21) and B (0.0024) observed in this correlation (Eq. 17) are greater than the corresponding A and B parameters observed for the Staudinger polymers prepared at 100 °C. This is interpreted to mean that polymerization conditions used by Boyer favored radical addition to DVB, more so than the corresponding conditions used by Staudinger. This implies that the proportion of extractable homopolymer produced by Boyer should be greater than that produced by Staudinger. This is indeed the case as indicated by relative percent extractable polymer reported by these investigators.

A more critical test of the gravimetric method for determination of crosslink density $(1/\lambda)$ would be a comparison of results obtained on the basis of Eq. 16 for poly(Sty-co-DVB) samples made via free-radical polymerization in various

Table 1. Correlation of observed crosslink density for poly(sty-*co*-DVB) with the corresponding observed carbon-13 line-width

Source	DVB Mole Fraction		S (ml/g)	line width (Hz)
	(y) *claimed*	(x) *observed*		
Rohm and Hass Co.	0.001	0.007 ± 0.002	8.14	15
Dow Chem. Co.	0.001	0.009 ± 0.002	7.19	33
Bio-Rad Labs. (SX-1)	0.01	0.010 ± 0.002	6.95	32
Rohm and Hass Co.	0.005	0.014 ± 0.003	5.82	49
Bio-Rad Labs (SX-2)	0.02	0.023 ± 0.002	4.13	250
Bio-Rad Labs (SX-3)	0.03	0.034 ± 0.002	3.02	320
Bio-Rad Labs (SX-4)	0.04	0.042 ± 0.002	2.44	360
Polysciences, Inc.	0.005	0.07 ± 0.01	1.65	620
Bio-Rad Labs (SX-8)	0.08	0.080 ± 0.002	1.48	900
Bio-Rad Labs (SX-12)	0.12	0.112 ± 0.003	0.94	2400

Claimed is the DVB mole fraction (y) in the monomer reaction mixture made to undergo polymerization by the manufacturer.
Observed is the average value for the DVB mole fraction (x) in the polymer as determined by swelling measurements (Eq. 16) in 10 different liquids (including $CHCl_3$).
S is mL of chloroform sorbed per gram of $(Sty)_{1-x}(DVB)_x$.

laboratories with corresponding results obtained using a method of proven reliability. Although such an alternative method does not yet exist, several investigators have reported [131–136] that high-resolution NMR spectra can be obtained on $(Sty)_{1-x}(DVB)_x$, and that the line-width in the ^{13}C spectra appear to correlate well with the respective reported DVB mole fractions, x, that were less than x = 0.06. We decided [137], therefore, to attempt correlation of the ^{13}C spectra linewidth observed for $DCCl_3$-swollen poly(Sty-*co*-DVB) samples obtained from four different commercial sources with the corresponding crosslink densities $[1/\lambda = x/(1 + x)]$ determined on the basis of the volume of liquid sorbed (S) using Eq. 16.

The commercial sources of these samples, the mole fraction (y) of DVB in the monomer mixture made to undergo polymerization [138], and the corresponding average crosslink density $(1/\lambda$, equal to approx. x when x is <0.1), determined on the basis of swelling in ten different liquids, are collected in Table 1, along with the corresponding linewidth $(v_{1/2})$ observed in the ^{13}C NMR spectra (Fig. 14).

The log/log plot of $v_{1/2}$ as a function of $1/\lambda$ (Fig. 15) show a linear relationship given by:

$$\log(v_{1/2}) = 1.6 \log(1/v) + 4.8 \tag{19}$$

or:

$$v_{1/2} = 6 \times 10^4 (1/\lambda)^{1.6}. \tag{19a}$$

Marchenkov and Khitrin [139] have performed a theoretical analysis relating $v_{1/2}$ to intramolecular dipole-dipole magnetic interactions. They claim $v_{1/2}$ should

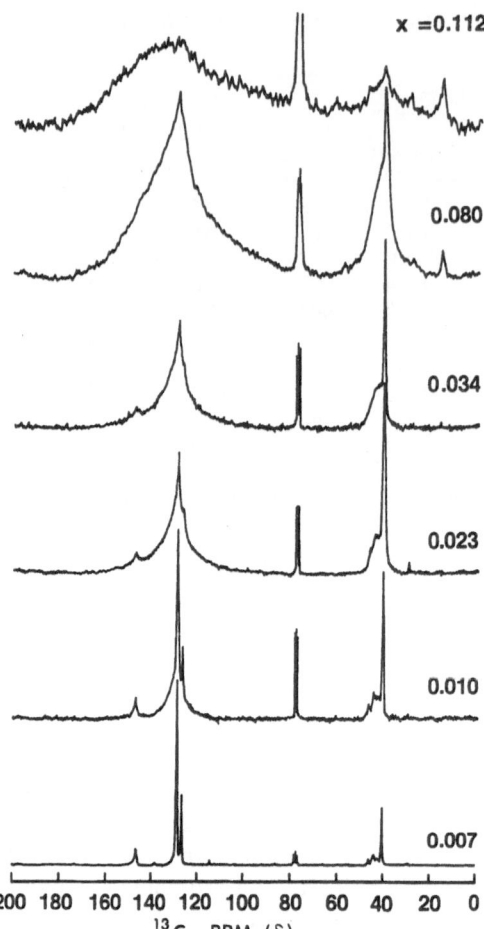

Fig. 14. 50-MHz ^{13}NMR spectra of $(Sty)_{1-x}(DVB)_x$ particles swelled to liquid saturation in CDCl$_3$, as a function of x

be proportional to the 4th root of $1/\lambda$ when, as in the case for divinylbenzene-crosslinked polymers, the length of the crosslinkages is less than the average chain length between crosslinks. The experimental observations reported here confirm a relation between $\nu_{1/2}$ and $1/\lambda$, but indicate the dependence on $1/\lambda$ is approximately the 3/2 power for divinylbenzene-crosslinked polystyrenes.

The data listed in Table 1 also confirm the observation made in the comparison of swelling data reported for polymers made in Staudinger's [1] and Boyer's [39] laboratories under ostensibly the same conditions, namely that the ratio of DVB mole fraction in the monomer charge (y) to the mole fraction of DVB in the resulting polymer (x) varies significantly from one laboratory to another, which presumably reflects small differences in experimental conditions, such as the electrophilicity of the initiating free radical(s) or the time profile of their concentration (or design of the equipment, such as the surface to volume ratios).

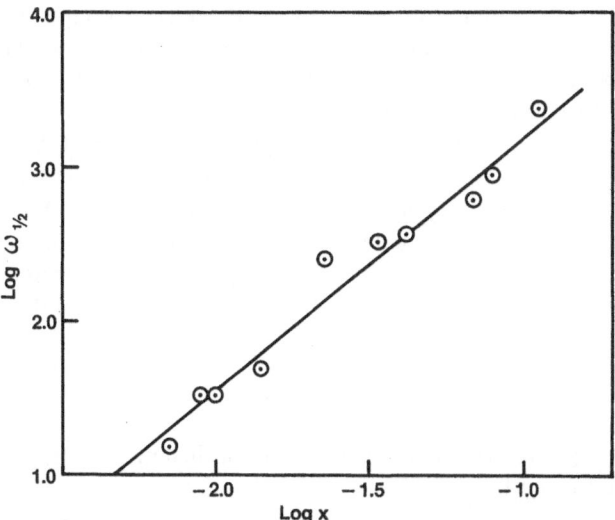

Fig. 15. Log-Log plot of aromatic carbon line-width ($\omega_{1/2}$) as a function of the DVB mole-fraction, x, in the $(Sty)_{1-x}(DVB)_x$ sample

3.2.4 Applicability to Other Polymers

3.2.4.1 Poly[block (Sty)-co-DVB]

The copolymers of styrene and DVB discussed so far were made via free radical polymerization, which is known to produce crosslinked polymer with a very broad distribution of λ in the polystyrene segments between cross-link junctions. On the other hand anionic polymerization of styrene monomer [140] to give the corresponding "living" dicarbanion n-mer[1] followed by addition of DVB monomer gives a crosslinked network with a very narrow range of λ (i.e. twice the number of monomer units) between "nodules" of $(DVB)_y$.

$$x(\cdot)\, CH_2CHPh(-) \rightarrow (-)(CHPhCH_2)_{x/2}\,(CH_2CHPh)_{x/2}\,(-)$$

$$(-)(Sty)_x(-) + yDVB \rightarrow (Sty)_x(DVB)_y .$$

Swelling of these polymers will depend not only on λ, but also on the ratio x/y, such that the parameters C and $\lambda_0^{1/3}$ (eq. 14) observed for these polymers in a given liquid should be characteristically different from those exhibited by polymers made via free-radical polymerization.

Swelling data for Sty-co-DVB polymers that have very-narrow-range molecular weight distributions for the poly(styrene) segments between poly(DVB) "nodules" have been reported by Rempp [141–143] and his coworkers. The size of these "nodules" and the number of polystyrene segments covalently bonded to a given

---

[1] n-mer refers to a short range degree of polymerisation to produce a polymer with two functional endgroups.

"nodule" was a function of the molar proportion of DVB added subsequently relative to the number of moles of available "living" carbanion polystyrene end groups at the time of the addition. Rempp reported his swelling data as the ratio of volume (V_g) of the gel to volume (V_0) of dry polymer, which was correlated with the molecular weight (M) of the polystyrene segments between $(DVB)_x$ "nodules". These data (Fig. 2 and Table 1 of Ref. 141, and Table 2 of Ref. 143) were re-calculated to give S, in ml/g of polymer, and the number of backbone carbon atoms (λ) in the polystyrene segments between "nodules". When the calculated values for S were plotted as a function of the calculated $\lambda^{1/3}$, straight lines were obtained as expected. The lines (Fig. 16) that represent the set of data published in 1970 (Table 2 of Ref. 143) are given by:

$$S = C(\lambda^{1/3} - 2.75),$$

but the lines (Fig. 17) that represent the two sets of data published in 1978 (Fig. 2 and Table 1 of Ref. 141) are given by:

$$S = C(\lambda^{1/3} - 5.0).$$

Why the constant $\lambda_0^{1/3}$ for Rempp's data [141] published in 1978 is twice that for the corresponding constant observed for his data published in 1970 is not understood. The $\lambda_0^{1/3}$ for the 1978 data, however, is unusually high relative to that observed for sty-*co*-DVB polymers made via free-radical polymerization (Figs. 9—11; $\lambda_0^{1/3} = 1.77 \pm 0.13$).

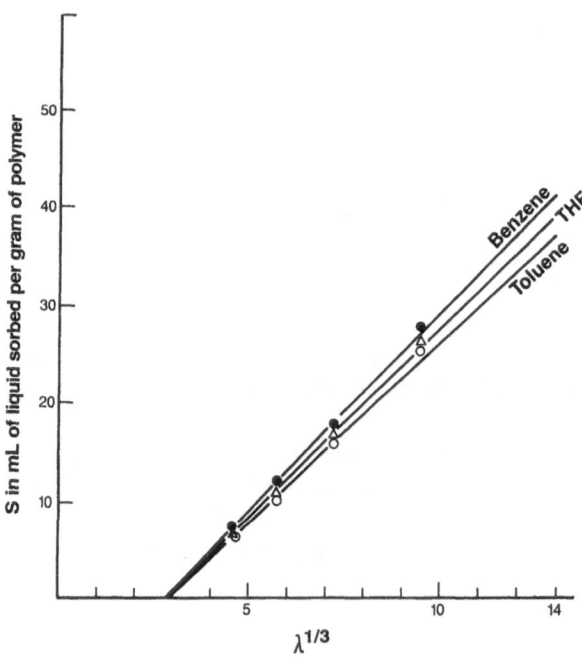

Fig. 16. Correlation of S in mL of sorbed liquid per gram of poly(Sty-*block*-DVB) (prepared via anion polymerization by Rempp [143] in 1970) with the corresponding calculated cube root of the number of backbone carbon atoms in the polystyrene segments between "nodules" of DVB

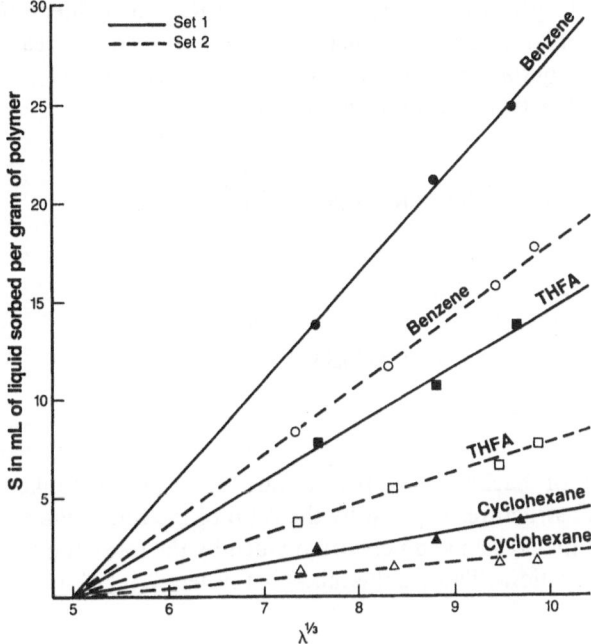

Fig. 17. Correlation of two sets of S, in mL of sorbed liquid per gram of poly(Sty-*block*-DVB) [prepared via anion polymerization by Rempp [141] in 1978] with the corresponding calculated cube root of the number of backbone carbon atoms in the polystyrene segments between "nodules" of DVB

The relative swelling powers of benzene (C_{1B} = 5.4), tetrahydrofurfuryl alcohol (C_{1THFA} = 2.9) and cyclohexane (C_{1cH} = 0.8) observed for the first set of co-polymers reported in 1978 (solid lines, Fig. 17) are uniformly about 1.8 fold greater that the corresponding relative swelling powers (C_{2B} = 3.5, C_{2THFA} = 1.6, C_{cH} = 0.4) calculated for the second set of copolymers reported in 1978 (dashed lines, Fig. 17). It is suspected that this difference may be attributable to the relative amount of DVB added to the available polystyrene carbanion end-groups to convert the "living" poly(styrene) polymers to the corresponding crosslinked polymer network. This rationale is consistent with the expectation that the swelling power of a given liquid, with respect to a given cross-linked polymer, will be to some degree a function of the molecular structure of the crosslinking "nodules", and that of the pendent groups, as well as of the number of carbon atoms in the backbone of the polymer segments between crosslink junctions.

3.2.4.2 "Macronet" Crosslinked Polyesters and Polyacrylates

Swelling data for "macronet" crosslinked polyesters were reported by Takahashi [144]. His polymers were prepared by causing a mixture of propylene glycol and an equivalent amount of diacid (maleic anhydride plus succinic acid) to undergo polycondensation at 220 °C to give the corresponding polyester. The centers of unsaturation in the soluble polymer obtained thereby were then made to react further via free-radical copolymerization with two equivalents of styrene to give the corresponding crosslinked network as indicated below:

$$\left[\left(O_2CCH=CHCO_2CH_2\overset{\overset{\displaystyle CH_3}{|}}{C}H\right)_x\left(O_2CCH_2CH_2CO_2CH_2\overset{\overset{\displaystyle CH_3}{|}}{C}H\right)_y\right]_n$$

$$\left[\left[\overset{\overset{\displaystyle |}{\underset{\underset{\displaystyle PhCHCH_2}{|}}{O_2CCHCHCO_2CHCH}}}{\underset{}{}}\overset{CH_2CHPh\ CH_3}{}\right]_x\left[O_2CCH_2CH_2CO_2CH_2\overset{\overset{\displaystyle CH_3}{|}}{C}H\right]_y\right]_n$$

The average number (λ) of backbone carbon atoms between crosslinked junctions in such "macronet" structures is given by the ratio of the total number of atoms in the backbones of the crosslinked network divided by the total number of atoms that form crosslink junctions. Thus, in this macronet system, λ is given by:

$$\lambda = (6x + 4y)/x,$$

where x and y are as defined in the crossliked polymer structure shown above. The sorption capacity, S, calculated from Takahashi's data (Table 1 of Ref. 144) was correlated with the corresponding $\lambda^{1/3}$, calculated from the data reported in Table 3 of Ref. 144. In every case, a linear relationship was obtained (Fig. 18) as given by:

$$S = C(\lambda^{1/3} - 2.05 \pm 0.05).$$

An approximately linear relationship of S as a function of $\lambda^{1/3}$ is also exhibited by hydrophilic "macronet" crosslinked polymers as indicated by the results deduced via a similar reconsideration of the swelling and molecular weight data reported by Refojo [145], who studied swelling of 2,3-dihydroxypropyl metha-crylate-co-tetraethylene glycol dimethacrylate polymers in water.

$$\left[\left[\overset{\overset{\displaystyle CH_3}{|}}{\underset{\underset{\underset{\displaystyle OCH_2CHCH_2OH}{|}}{\underset{\displaystyle C=O\ OH}{|}}}{CH_2C}}\right]_x\left[\overset{\overset{\displaystyle CH_3}{|}}{\underset{\underset{\displaystyle O}{\|}}{CH_2C}}-C-O-(CH_2CHO)_4\overset{\overset{\displaystyle CH_3}{|}\ \overset{\displaystyle O}{\|}}{\underset{\underset{\displaystyle CH_3}{|}}{C}-C}-CH_2\right]_y\right]_n$$

In these crosslinked macronet copolymers, λ is given by:

$$\lambda = (2x + 19y)/2y.$$

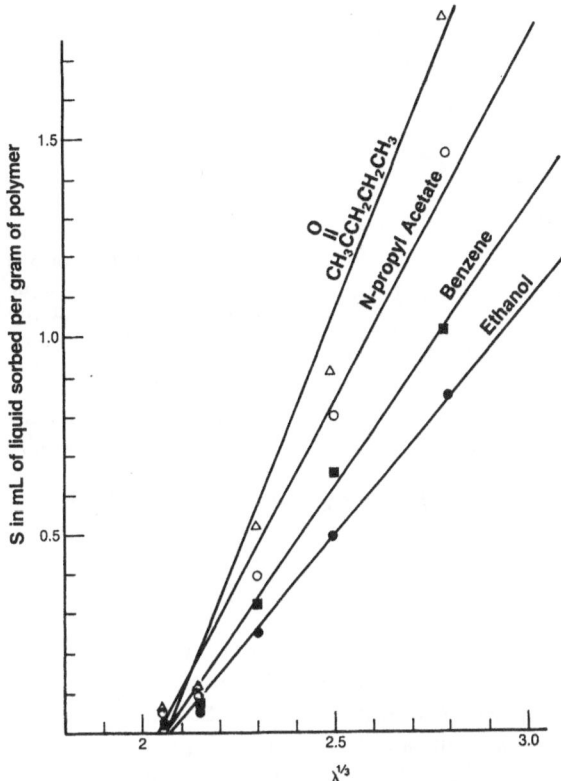

Fig. 18. Correlation of S, in mL of sorbed liquid per gram of crosslinked macronet polyester (prepared by Takahashi [144] in 1983) with the corresponding cube root of the average number of carbon atoms between crosslink junctions

The correlation of calculated S with calculated $\lambda^{1/3}$ (Fig. 19) show that all but one of the data points fall on the line given by:

$$S = 1.08(\lambda^{1/3} - 1).$$

It is suspected, however, that deviation from linearity at high mole fraction of the comonomer may be the rule for "macronet" crosslinked polymers of the type X-co-Y, because the observed S is a function of two variables that change with (1) the mole-fraction of monomer Y, which provides the crosslinking junction, i.e. $1/\lambda$ and (2) the relative affinity of the solvent for the components X and Y. Thus, the relatively low S at $\lambda^{1/3} = 4.78$ in Fig. 19 may reflect a lower affinity of water for the 2,3-dihydroxypropyl methacrylate (component X) relative to the corresponding affinity for tetraethylene glycol di-methacrylate (component Y), the cumulative effect of which can be significantly larger at high X/Y relative to that at low X/Y.

The above results indicate that Eq. 14 may be general for liquid sorption by crosslinked polymer networks, i.e. the volume, S, of liquid sorbed per gram of crosslinked polymer increases linearly with the cube root of the average number, λ, of backbone atoms in the segments between crosslink junctions. The constants

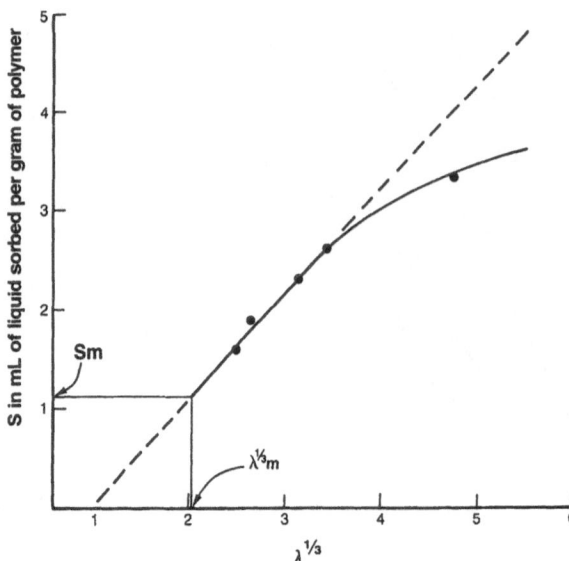

Fig. 19. Correlation of S in mL of water sorbed at liquid saturation per gram of Gly-MA-*co*-TEGDMA copolymers (prepared by Refojo [145] in 1965) with the corresponding calculated cube root of the average number of backbone carbon atoms between crosslink junctions

$\lambda_0^{1/3}$ and C, however, are characteristic of the polymer-liquid system. Once $\lambda_0^{1/3}$ for a given type of crosslinked copolymer is established, however, the observed S at gel-saturation in a liquid with known swelling power, C, can be used to establish the crosslink density $(1/\lambda)$ of an unknown sample within the type of crosslinked polymer class (Eq. 16). Alternatively, the observed S for sorption by a given copolymer network with known λ_0 can be used to establish the relative swelling power of test-liquids (Eq. 14). Correlation of such data with the molecular structures of the test-liquids that comprise homologous series should provide insight into the electronic and steric factors that affect molecular sorption, which is important not only for polymer swelling and polymer permeation, but also in the understanding of solvent effects which influence the kinetics of organic reactions in solution.

3.3 Concept of an Adsorption Parameter, α_s

3.3.1 Sorption Model

In the Theoretical Considerations (Sect. 3.2.1), of polymer swelling, it was stated that the number, α_s, of *ad*sorbed molecules per accessible monomer unit of crosslinked polymer at liquid-saturation (Eq. 15) can be calculated directly from the observed swelling power, C, as defined in Eq. 14. It follows, therefore, that Eq. 14 can be re-written in the form:

$$\alpha_G = \alpha_s(\lambda^{1/3} - \lambda_0^{1/3}) = \alpha_s\Lambda, \tag{20}$$

where λ and λ_0 are as defined previously (Eq. 14; Sect. 3.2.1) such that the difference, $(\lambda^{1/3} - \lambda_0^{1/3})$ or Λ, is a dimensionless number that reflects the "looseness" of the macrostructural crosslinked network, and the product of α_s and Λ, i.e. α_G, is the total number of sorbed molecules per monomer unit of polymer at liquid-saturation, both *ad*sorbed and non-*ad*sorbed.

The protocol for the determination of α requires that the set of film samples comprised of $(Sty)_{1-x}(DVB)_x$ particles enmeshed in PTFE fibers (Figs. 7C and 7D) be allowed to swell to saturation as described in Sect. 3.1. The distribution of sorbed liquid in such composite films at the saturated state is illustrated schematically in Fig. 20, which indicates two types of sorbed molecules, as required by Eq. 20, i.e. those ($\bigcirc$) that are immobilized by adsorption to the polymer molecules, and those ($\blacksquare$ and $\triangle$) that are not adsorbed as described in the caption for Fig. 20.

Although the non-adsorbed molecules ($\blacksquare$) within the liquid-saturated particles are in exchange equilibrium with the adsorbed molecules ($\bigcirc$), the number, α_s, of adsorbed molecules per monomer unit of polymer remains constant. Thus, α_s can be considered to be the average dynamic packing density of sorbed molecules *ad*sorbed to a monomer unit of polymer at liquid-saturation, and the magnitude of α_s should (1) vary directly with the force of attraction between

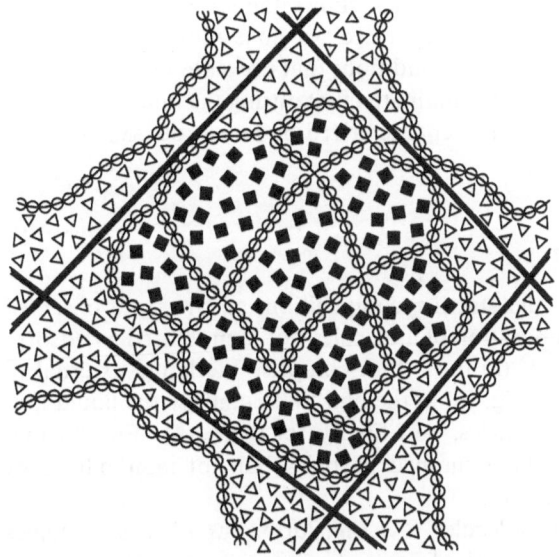

Fig. 20. Schematic representation of a composite membrane (Figs. 1 and 7) at liquid saturation showing a single gelled particle enmeshed in PTFE microfibers as described in the text. The *bold straight lines* represent the PTFE fibers. The entangled network of *curved lines* represent the crosslinked polymer that supports the liquid saturated gel. Each *empty circle* ($\bigcirc$), superimposed on the curvy lines, represents a set of molecules (α_s, as defined in Eq. 20) adsorbed to an accessible monomer unit. The *filled squares* ($\blacksquare$) represent liquid molecules that are sorbed by the gelled particles, but not immobilized by adsorption to the polymer molecules. The *empty triangles* ($\triangle$) represent liquid molecules that surround the liquid saturated gel particles enmeshed in the composite membrane. The excess liquid, in contact with the external surface of the liquid saturated composite membrane, is not shown

the sorbed molecule and the monomer unit of the sorbent polymer, and (2) vary inversely with the opposing force due to steric hindrance, as expressed relative to such volume exclusion limitations as are imposed by the finite adsorption area available on all sides of a monomer unit. This implies that the magnitude of α_G (Eq. 20) will depend upon how well the molecular structure of the sorbed liquid can be accommodated three-dimensionally by the molecular structure of a monomer unit of the polymer at liquid-saturation, which means that it should be possible to correlate the magnitude of swelling with the molecular structures of sorbed liquid and sorbent polymer. Before undertaking such a study, however, it is necessary to test the validity of the sorption model as pictured in Fig. 20.

3.3.2 Verification of the Model

If there are indeed two types of sorbed molecules, as implied by Fig. 20 and Eq. 20, then it should be possible to detect the difference in retentivity as the system is evaporated from liquid-saturation to virtual dryness. We reported [146–149] that this is indeed the case, based on numerous time-studies of evaporation using acetone, chloroform or toluene as the volatile sorbed liquid. A typical example of such a time-study is recorded in Fig. 21, which shows the sequential changes in kinetics as the (in this case $CHCl_3$) liquid-saturated system is allowed to evaporate to dryness at constant temperature under conditions that prevent sample shrinkage [148]. The non-adsorbed molecules are eliminated first during the interval required for α_t to decrease from α_0 to α'_s, which is characterized by zero-order kinetics [146, 147] (Inset Fig. 21), i.e.:

$$\alpha_t = \alpha_0 - rt, \tag{21}$$

where α_0 (some value $> \alpha'_s$) is the starting composition, and r is the observed zero-order rate-constant for evaporation of molecules not immobilized by adsorption [i.e. sequential elimination of (a) those that comprise the liquid external to the composite film, (b) those ($\triangle$; Fig. 20) in the film that comprise the liquid that surround the liquid-saturated particles, and finally (c) those ($\blacksquare$; Fig. 20) that comprise the liquid sorbed by the enmeshed particles but not immobilized by adsorption to the polymer molecules.

Elimination of the adsorbed molecules ($\bigcirc$, Fig. 20) begins when α_t becomes equal to α'_s. During the interval required for α_t to decrease from α'_s to α'_g, the kinetics of desorption is first-order with respect to α_t, i.e.

$$\alpha_t = \alpha'_s e^{-k'}(t - t's), \tag{22}$$

where k' is the first-order rate constant for desorption from poly(Sty-co-DVB) in the rubbery state, and t'_s is the time that marks incipient desorption of those molecules ($\bigcirc$) immobilized by adsorption to monomer units of polymer in the rubbery state.

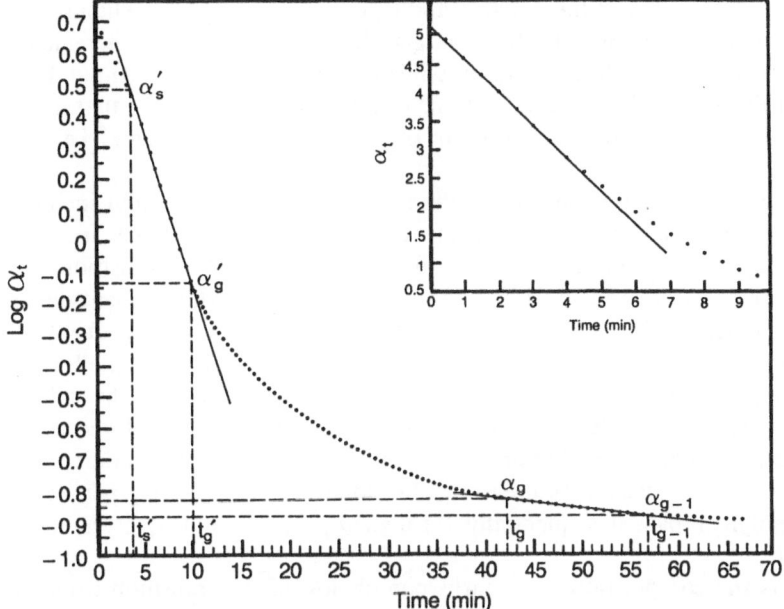

Fig. 21. A typical example of many time-studies of desorption from liquid saturated poly(Sty-*co*-DVB) particles enmeshed in PTFE microfibers. The *Inset* records the number, α_t, of residual sorbed molecules [in this case $CHCl_3$ in $(Sty)_{98}(DVB)_2$] per phenyl group of polymer over the first 10 min of desorption at 23 °C. The main plot records the logarithm of α_t over the first 70 min of the time-study. The "breakpoints" in the kinetics of desorption at α'_s, α'_g and α_g mark sequentially the compositions at incipient elimination of molecules immobilized by adsorption, incipient transition from the rubbery state of the system to the glassy state, and finally completion of this transition as discussed in the text

Termination of first-order kinetics, when α_t becomes equal to α'_g, signals incipient transition from the rubbery state to the glassy state. Restoration of first-order kinetics, when α_t becomes equal to α_g, signals completion of this transition. Thereafter, α_t is given by a linear combination of exponential decay functions:

$$\alpha_t = \alpha_g \sum_{i=1}^{n} f_i \, e^{-k_i(t-t_g)}, \tag{23}$$

where n is the number of populations of different molecular environments (not more than 6) [148, 149] created during transition from the rubbery state to the glassy state, $f_i = \alpha_{g,i}/\alpha_g$ is the fraction of α_g trapped in the *i*th population, and k_i is the first-order rate constant for decay of that population, and i is the numerical sequence of decreasing k.

The rate constant k_1 for the population with the fastest decay rate was shown to be equal to the rate constant, k', for desorption from polymer in the rubbery state [148], and the rate constants k_i for the set of n populations trapped in the glassy state (attained when α_t becomes equal to α_g) are related one to another by:

$$\text{Log } k_i = \text{Log } k_0 - mi, \tag{24}$$

where m is characteristic of the polymer and k_0 is characteristic of the liquid [148, 149], such that k_1 is about 10^5 times larger than k_6. Because the incremental decrease in k_i is relatively large, it was possible to separate each exponential decay function in Eq. 23 and thereby establish the residual composition α_{g-i} that marks depletion of each successive population in the sequence of decreasing k_i, i.e. α_t at α_g, α_{g-1}, α_{g-2} etc.

That α_s', α_g', and α_g are the compositions that mark (sequentially) incipient elimination of *ad*sorbed molecules, incipient transition of the polymer from the rubbery state to the glassy state, and finally completion of this transition, was verified spectrophotometrically [150–152] using (a) p-dimethylaminobenzylidene malononitrile [153, 154] as an intramolecular-rotor fluorescence probe [150] (Fig. 22), (b) 2,2,6,6-tetramethyl-piperidine-1-oxyl (i.e. TEMPO) or 4-oxo-TEMPO as an EPR probe [151] (Figs. 23 and 24), and (c) proton NMR of the volatile liquid [152] (Figs. 25 and 26), the various spectral signals being monitored as the volatile component evaporated to dryness. In all cases sharp changes in the spectra occurred as expected when the number, α_t, of residual sorbed molecules per phenyl group of polymer passed sequentially through α_G, α_s', α_g', and α_g as noted in Figs. 22–26.

That α_g is the composition that marks completion of the transition from the rubbery state to the glassy state at $23.5 \pm 0.5\,^\circ$C is also supported by the good agreement with the corresponding composition reported by others [155–159], who

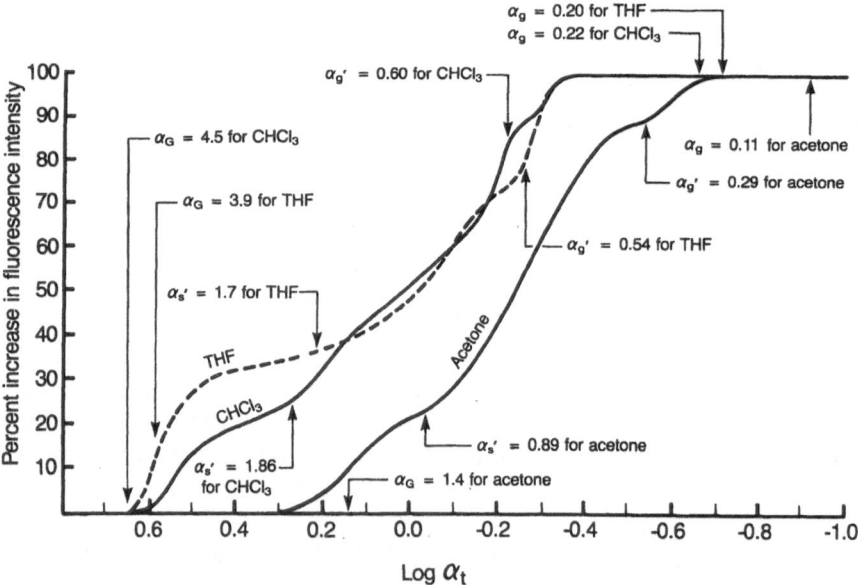

Fig. 22. Correlation of percent increase in fluorescence intensity emitted by a rotor-fluorescent probe molecule (p-dimethylaminobenzylidene malononitrile) with the corresponding logarithm of the number, α_t, of residual sorbed molecules (acetone, $CHCl_3$, or THF) per phenyl group of $(Sty)_{97}(DVB)_3$ as described in the text

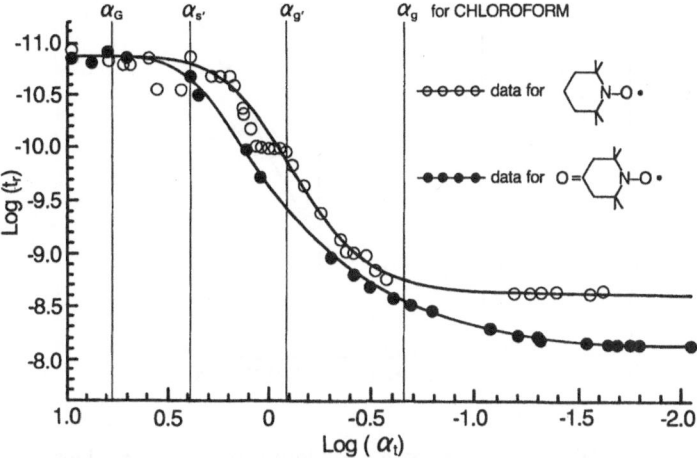

Fig. 23. Plot of the rotational correlation time (t_r) for nitroxide spin-probes (TEMPO and 4-oxo-TEMPO), in chloroform swollen $(Sty)_{98}(DVB)_2$ versus the logarithm of the corresponding residual number, α_t, of sorbed $CHCl_3$ per phenyl group of polymer

used thermodynamic methods to establish the glass-transition temperature of two-component polystyrene systems as a function of the weight percent of the second component.

Multireplicated time-studies that monitored desorption from liquid-saturated poly(Sty-*co*-DVB) to virtual dryness, using acetone ($\alpha_s = 0.92$), toluene ($\alpha_s = 1.96$) and chloroform ($\alpha_s = 3.00$) as sorbed liquids, showed [149] that the average composition at each successive breakpoint that signalled sequentially α'_s, α'_g, and

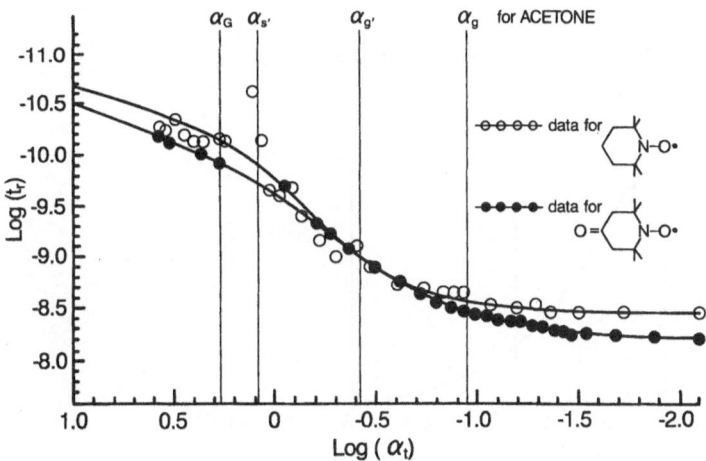

Fig. 24. Plot of the rotational correlation time (t_r) for TEMPO and 4-oxo-TEMPO, in acetone swollen $(Sty)_{98}(DVB)_2$ versus the logarithm of the corresponding number, α_t, of residual sorbed acetone molecules per phenyl group of polymer

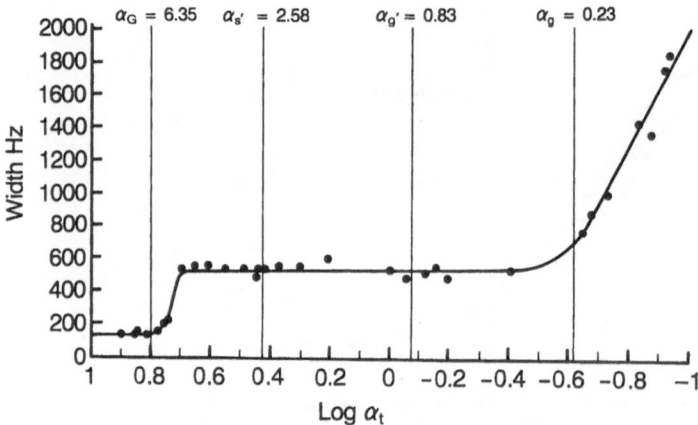

Fig. 25. Correlation of ^{1}H NMR Linewidth in Hz for dichloromethane with the logarithm of the corresponding number, α_t, of residual dichloromethane molecules per phenyl group of $(Sty)_{98}(DVB)_2$

α_g (Fig. 21) are given by:

$$\alpha_s' = 0.31\Lambda(\alpha_s + 1), \tag{25}$$

$$\alpha_g' = 0.10\Lambda(\alpha_s + 1), \tag{26}$$

$$\alpha_g = 0.055(\alpha_s + 1), \tag{27}$$

where Λ is the difference $(\lambda^{1/3} - \lambda_0^{1/3})$ as defined in Eqs. 14 and 20.

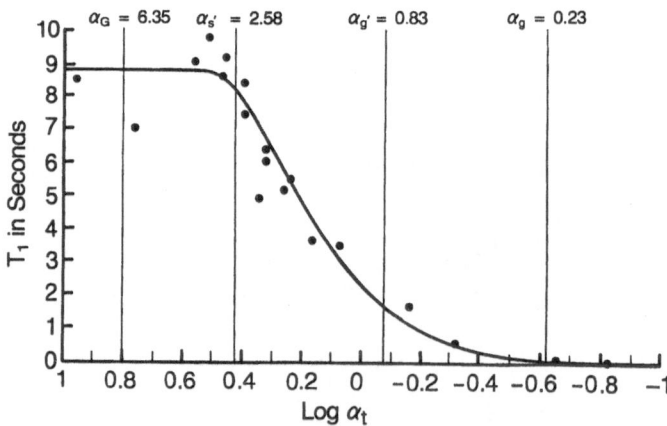

Fig. 26. Correlation of T_1 relaxation time (^{1}H NMR) in seconds with the logarithm of the corresponding number, α_t, of residual dichloromethane molecules per phenyl group of $(Sty)_{98}(DVB)_2$

These results (Eqs. 25–27) show that the residual composition α_t at each breakpoint in the kinetics of desorption, which marks some form of physical change in the system, is proportional to α_s (Eq. 20), i.e. the number of adsorbed molecules per accessible phenyl group of $(Sty)_{1-x}(DVB)_x$ at liquid-saturation, or in "true solution" when x is zero. Moreover the residual compositions (α_{g-1}, α_{g-2} etc.) in the glassy state, which mark the sequential depletions of populations present when α_t became α_g, are also related linearly [149] to α_s as expressed by:

$$\alpha_{g-i} = 0.055 A_{g-i}(\alpha_s + 1), \tag{28}$$

where the proportionality constant A_{g-i} after removal of the ith population is:

$$A_{g-i} = (1 - 0.18i). \tag{29}$$

Since the relative concentration of residual adsorbed molecules at every level that can be identified by some physical measurement (from saturation to virtual dryness) is a linear function of α_s, it is reasonable to conclude that α_s might indeed be a parameter of fundamental scientific value, potentially helpful in improving our understanding of phenomena that involve molecular association, e.g. chromatography, membrane permeation (Sect. 2.1), solubility and thermally induced gel-formation.

3.4 Correlation of α with Molecular Structures

If one accepts that the number of adsorbed molecules per accessible phenyl group in poly(Sty-co-DVB) at liquid-saturation (i.e. α_s, hereafter in this section referred to simply as α without the subscript s) is an adsorption parameter that reflects how well the molecular structure of the adsorbed molecule is accommodated by the molecular structure of the repeat unit, then it follows that α should vary in accordance with the electronic nature and molecular architecture of the adsorbed species. That this is indeed the case is made evident by the α data collected for homologous series of liquids ZR in which Z [a substituent that has relatively strong affinity for the phenyl groups in poly(Sty-co-DVB)] is kept constant, while R [the remaining molecular structure] is varied systematically [160–164].

3.4.1 Substituted Benzenes

The mono- and di-substituted benzenes, for which the relative swelling power (C as defined in Eq. 14) have been determined thus far, are listed in Tables 2–4. In every case the volume (S) of sorbed liquid per gram of polymer increased linearly with $\lambda^{1/3}$ as noted in Figs. 9–11, and the square of the correlation coefficient (r) to the line of best fit, determined by linear regression analysis of the set of six S-data points, was in every case $r^2 > 0.99$, and $\lambda_0^{1/3}$ was equal to 1.84 ± 0.09. The corresponding C and α (calculated therefrom; Eq. 15) are also collected in Tables 2–4. These data show clearly that the value α = 2.50 for benzene is greater *than that* for any benzene derivative listed therein. Any substituent in place of hydrogen on the benzene ring structure serves to decrease α, regardless of its perceived electronic contribution to the force of mutual attraction, which implies that the group's negative contribution owing to steric hindrance is greater than any positive contribution owing to the electronic interaction of the substituent with the phenyl ring. The magnitude of decrease observed with electron-withdrawing substituents (CF_3, NO_2), however, is greater than that observed with electron-donating substituents (CH_3, $N(CH_3)_2$) of approximately the same size (Table 2). The order of α in a given series, therefore, reflects the net result of the positive or negative contribution to electronic attraction and the negative effect of steric hindrance imparted by the substituent. Thus, α for monosubstituted benzenes (Table 2) falls in the order $H > Cl > F > Br > I$ and $H > CH_3 > CCl_3 > CF_3$, and in the order $CH_3 > CH_2CH_3 > CH(CH_3)_2 > C(CH_3)_3$.

Table 2. Parameters for polystyrene-liquid systems: Monosubstituted benzene liquids: $\delta_{pol} = 9.5$ $(cal/mL)^{1/2}$

Liquid	C	α	$\alpha/2.5$	δ	χ
PhH	2.14	2.50	1.00	9.16	0.195
PhCl	2.18	2.23	0.89	9.67	0.170
PhF	1.98	2.19	0.88	9.11	0.292
PhBr	2.06	2.04	0.82	9.87	0.243
PhI	1.88	1.75	0.70	10.13	0.353
$PhCH_3$	2.02	1.98	0.79	8.93	0.268
$PhCCl_3$	2.00	1.66	0.66		0.280
$PhCF_3$	0.71	0.59	0.24	8.22	1.07
$PhOCH_3$	2.00	1.92	0.77		0.280
$PhN(CH_3)_2$	2.25	1.84	0.74	9.84	0.128
$PhNH_2$	1.49	1.70	0.68	10.42	0.591
$PhNO_2$	1.57	1.14	0.46		0.542
$PhCH_2CH_3$	1.84	1.55	0.62	8.84	0.378
$PhCH(CH_3)_2$	1.65	1.23	0.49	8.60	0.494
$PhC(CH_3)_3$	1.47	0.99	0.40		0.603
Ph-cyclohexane	1.80	1.11	0.46		0.402
$PhCH_2OH$	0.82	0.83	0.33		1.00
$PhCH_2Cl$	2.39	2.16	0.86		0.042
$PhCOCH_3$	1.88	1.68	0.67	9.68	0.353
$PhCH_2O_2CCH_3$	1.94	1.26	0.50		0.317
$PhCH_2OCH_2CH_3$	1.85	1.34	0.54		0.371
$PhCH_2Ph$	2.07	1.29	0.52		0.237
Pyridine	1.63	2.11	0.84		0.506

C is the relative swelling power (in mL/g) of the liquid as defined in Eq. 14.
α is the adsorption parameter (i.e. the number of adsorbed molecules per phenyl group of polymer at liquid saturation) as defined in Eq. 15.
$\alpha/2.5$ is the ratio of $\alpha_{liquid}/\alpha_{PhH}$.
δ is the solubility parameter in $(cal/mL)^{1/2}$ as reported by Hoy [34] and by Hansen [176].
χ_1 is the Flory-Huggins interaction parameter χ at $v = 1$, where v is the volume fraction of polymer in the polystyrene-liquid system. χ_1 was calculated from the corresponding observed C using Eq. 45 in Sect. 4.2. χ_v at any other level of v can be calculated from χ_1 by means of Eq. 44.

Table 3. Parameters for polystyrene-liquid systems: Disubstituted benzene liquids: $\delta_{pol} = 9.5$

Liquid	C	α	$\alpha/2.5$	δ	χ
Tetrahydronaphthalene	2.22	1.69	0.68	9.50a	0.146
o-xylene	2.02	1.74	0.70	9.06a	0.268
m-xylene	1.80	1.53	0.62	8.88a	0.402
p-xylene	1.73	1.46	0.58	8.83a	0.445
o-chlorotoluene	1.99	1.77	0.71	10.24b	0.286
m-chlorotoluene	1.91	1.68	0.67	10.24b	0.335
p-chlorotoluene	1.81	1.59	0.64	10.24b	0.396
o-dichlorobenzene	1.95	1.80	0.72	10.04a	0.311
m-dichlorobenzene	2.05	1.87	0.75	9.80a	0.250

C, α, $\alpha/2.5$, δ, and χ_1 are as defined in Table 2.
(a) δ data reported by Hoy [34] and/or Hansen [176].
(b) δ value calculated by the method of component contributions in the manner outlined by Van Krevelen [27].

The decrease in α resulting from hydrogen bonding is also readily apparent [160] when $\alpha = 1.70$ for $PhNH_2$ is compared to $\alpha = 1.84$ for $PhN(CH_3)_2$ (Table 2), and $\alpha = 0.83$ for $PhCH_2OH$ is compared with $\alpha = 1.26$ and 1.34 for $PhCH_2OAc$ and $PhCH_2OEt$ respectively (Table 4). Self-association of the liquid by hydrogen bonding increases the "apparent" size of the substituent attached to the portion of the molecule being accommodated by the phenyl group in the polymer, resulting in increased steric hindrance.

Table 4. Parameters for polystyrene-liquid systems: phenyl $(CH_2)_nH$ Liquids: $\delta_{pol} = 9.5$

n	C_n	α_n	δ_n	$(9.5 - \delta_n)^2$	χ_1
0	2.14	2.50	9.16a	0.12	0.195
1	2.00	1.98	8.93a	0.32	0.280
2	1.84	1.55	8.84a	0.44	0.378
3	1.78	1.33	8.64a	0.74	0.414
4	1.69	1.13	8.58a	0.85	0.469
5	1.53	0.93	8.50b	1.00	0.567
6	1.45	0.80	8.44b	1.12	0.616
7	1.30	0.66	8.39b, c	1.23	0.707
8	1.07	0.50	8.35b, c	1.32	0.847
9	0.92	0.40	8.31b, c	1.42	0.939
10	0.74	0.30	8.28b, c	1.49	1.05

C, α, and χ_1 are as defined in the footnotes of Table 2.
δ_n is the solubility parameter in $(cal/mL)^{1/2}$
(a) data reported by Hoy [34] and/or by Hansen [176],
(b) data calculated by the method of component contributions in the manner outlined by Van Krevelen [27].
9.5 is δ_{pol} with respect to benzene substituted liquids.

Similarly, any positive electronic contribution by a second substituent on the aromatic ring is more than offset by its negative contribution owing to steric hindrance [160]. Thus, α for toluene is greater than for any of the xylenes (Table 3), and the order for chloro-substituted benzenes and toluenes is PhH > PhCl > PhCH$_3$ > ClPhCl > ClPhCH$_3$ > CH$_3$PhCH$_3$ (Table 3). The data for the xylenes and the chlorotoluenes show that the corresponding α are in the order *ortho* > *meta* > *para*, which again indicates how steric considerations can override the effect of electronic contribution.

The above results emphasize that the importance of steric hindrance relative to electronic attraction is much greater in the case of molecular association with polymers than it is in classical physico-organic chemistry regarding reactions of small molecules in solution [160]. The reason of course is that the attractive forces involved in molecular association are much weaker than the short-range forces of electronic interaction (at binding distances) involved in organic reactions.

In summary the results observed in these studies [160] of poly(Sty-*co*-DVB) swelling in aromatic liquids serve to show that the method of measuring α is so sensitive that it can detect an effect caused by even the smallest modification in the molecular geometry of attached substituents, and that these differences correlate qualitatively with expectation based on the known principles of physico-organic chemistry of aromatic compounds. Since the observed α is the net effect of electronic attraction and steric hindrance between the sorbed molecule and the adsorption site, i.e. the monomer unit of the polymer, it would be impossible to separate quantitatively the electronic and steric contributions of a particular substituent. The ability to make such a differentiation, however, appears to be more promising with liquids that comprise homologous series of the type Z(CH$_2$)$_n$H (where Z is a phenyl, chloro, bromo or iodo substituent), since the added electronic contribution to Z by each additional methylene group is well known to be extremely small when n becomes >3 [165].

3.4.2 Alpha Substituted Acyclic Alkanes

3.4.2.1 Linear Alkanes

The test-liquids used in these studies of Z(CH$_2$)$_n$H sorption by poly(Sty-*co*-DVB) are listed in Tables 4–7. In every case, the volume (S) of test-liquid sorbed per gram of polymer increased linearly with $\lambda^{1/3}$ of that polymer (in accordance with Eq. 14 and the examples shown in Fig. 27); the intercept $\lambda_0^{1/3}$ was 1.77 ± 0.14; and the corresponding r^2 was >0.99 as indicated by the examples recorded in Fig. 27. The parameters C and α for these liquids are also collected in Tables 4–7. It was noted [160–164] that each correlation of C_n with n for a given Z(CH$_2$)$_n$H series exhibits a unique pattern, even in a qualitative sense. Thus, when Z is phenyl (Fig. 28), C decreases linearly with n until n becomes 7, and thereafter deviates negatively [160] from the line established for the liquids with n < 8; when Z is iodo (Fig. 29), C begins to deviate negatively from linearity at n = 4. In fact the first four data points exhibit a small odd-even alternation [162] above and below a line given by 2.12 − 0.044n; when Z is bromo (Fig. 30), C increases to a maximal value [164] at n = 3 and then follows a pattern similar to that exhibited by the

Table 5. Parameters for polystyrene-liquid systems: $I(CH_2)_nH$ liquids: $\delta_{pol} = 10.1$

n	C_n	α_n	δ_n	$(10.1 - \delta_n)^2$	χ_1
1	2.08	3.47	10.1	0.0	0.231
2	2.03	2.64	9.69	0.17	0.262
3	2.03	2.17	9.36	0.54	0.262
4	1.94	1.78	9.15	0.90	0.317
5	1.82	1.45	8.98	1.25	0.390
6	1.72	1.22	8.84	1.59	0.451
7	1.57	1.00	8.74	1.85	0.542
8	1.36	0.78	8.65a	2.10	0.570
9	1.10	0.67	8.58a	2.31	0.829
10	0.89	0.43	8.51a	2.52	0.957
11	0.57	0.26	8.45a	2.72	1.15
12	0.35	0.15	8.43a	2.79	1.29

C, α, and χ_1 are as defined in the footnotes of Table 2.
δ_n is the solubility parameter in $(cal/mL)^{1/2}$, which were calculated by the method of component contributions in the manner outlined by Van Krevelen [27].
(a) The correlation (Fig. 57) of Log α with $(10.1 - \delta_n)^2$ indicates that this calculated value for δ_n may be too high (see Table 18).
10.1 is δ_{pol}, the solubility parameter of the polymer with respect to $I(CH_2)_nH$ liquids.

$Ph(CH_2)_nH$ liquids with $n > 3$ (Fig. 28). Since methyl bromide is not a liquid at room temperature and atmospheric pressure (the conditions chosen for measuring polymer swelling), C_1 (empty circle in Fig. 30) was not determined experimentally but rather it was obtained by extrapolation of C_n for bromoform and methylenebromide and of Log α_n for the $Br(CH_2)_nH$ liquids to methyl bromide as described in Ref. 164.

Table 6. Parameters for polystyrene-liquid systems: $Br(CH_2)_nH$ liquids: $\delta_{pol} = 10.5$

n	C_n	α_n	δ_n	$(10.5 - \delta_n)^2$	χ_1
1	[1.34]	[2.72]	10.85	0.20	0.683
2	1.62	2.26	9.81	0.48	0.512
3	1.79	2.05	9.60	0.81	0.408
4	1.73	1.68	9.42	1.17	0.445
5	1.67	1.44	9.29	1.46	0.481
6	1.64	1.22	9.21	1.66	0.500
7	1.60	1.06	9.13	1.88	0.524
8	1.46	0.87	9.06a	2.07a	0.609
9	1.31	0.71	9.01a	2.22a	0.701
10	1.06	0.53	8.97a	2.32a	0.853
11	0.87	0.41	8.94a	2.43a	0.969
12	0.64	0.28	8.89a	2.59a	1.11

C, α, δ, and χ_1 are as defined in the footnotes of Table 5.
10.5 is δ_{pol}, the solubility parameter in $(cal/mL)^{1/2}$ of the polymer with respect to $Br(CH_2)_nH$ liquids.
(a) The correlation of Log α_n with $(10.5 - \delta_n)^2$ (Fig. 58) indicates that this value calculated for δ_n may be too high (see Table 18).

Table 7. Parameters for polystyrene-liquid systems: $Cl(CH_2)_nH$ liquids: $\delta_{pol} = 10.0$

n	C_n	α_n	δ_n	$(10.0 - \delta_n)^2$	χ_1
1	[1.70]	[3.21]	10.30	0.09	0.463
2	[1.54]	[2.23]	9.70	0.09	0.551
3	1.63	1.94	9.17	0.69	0.506
4	1.42	1.55	8.98 (8.37a)	1.04	0.634
5	1.50	1.29	8.83	1.36	0.585
6	1.37	1.04	8.72	1.64	0.664
7	1.42	0.96	8.65	1.82	0.634
8	1.27	0.78	8.55	2.10	0.725
9	1.03	0.57	8.48b	2.31b	0.872
10	0.85	0.44	8.43b	2.46b	0.982
11	[0.60]	[0.29]	8.40b	2.59b	1.13
12	0.40	0.18	8.35b	2.72b	1.26

C, α, and χ_1 are as defined in the footnotes of Table 5.
δ_n calculated by the method of component contributions in the manner outlined by Van Krevelen [27].
(a) δ_4 reported by Hoy [34].
(b) The correlation of Log α_n with $(10.0 - \delta_n)^2$ (Fig. 59) indicates that this value calculated for δ_n may be too high (see Table 18).
10.0 is δ_{pol}, the solubility parameter in $(cal/mL)^{1/2}$ of the polymer with respect to $Cl(CH_2)_nH$ liquids.

The most interesting of the four series is exhibited by the $Cl(CH_2)_nH$ series [163]; the data for the liquids with $n < 8$ of this series (Fig. 31) exhibit a pronounced odd-even alternation above and below a line given by: $C_n = 1.67 - 0.04n$. The C_n data for liquids with $n > 7$ deviate negatively from linearity in the manner exhibited by the other series with phenyl, iodo or bromo in the 1 position (Figs. 29–30).

The relative swelling powers, $C_1 = 1.70$ and $C_2 = 1.54$ (Table 7) for methyl and ethyl chlorides were not determined experimentally because these two lowest members of the $Cl(CH_2)_nH$ series are gases under the conditions chosen to measure polymer swelling. Instead these values were deduced by extrapolation [163] of the relative swelling power data observed for $CHCl_3$ and CH_2Cl_2 (Table 8 of Ref. 161) to that reported here for CH_3Cl (Table 7) and by extrapolation of those values observed for Cl_2CHCH_2Cl and $ClCH_2CH_2Cl$ (Table 8 of Ref. 161) to that reported here for CH_3CH_2Cl (empty circle, Fig. 31). The general validity of such extrapolations is supported by similar relationships observed in ongoing studies of the relative swelling power, C_x, as a function of x in homologous series of the type $CH_{(4-x)}X_x$ and $C_2H_{(6-x)}X_x$, which will be reported fully at a later date.

The odd-even alternations displayed in half of the above studies (Figs. 29 and 31) are not yet fully understood. It is well known [166] however, that the molar refractions of $Z(CH_2)_nH$ liquids exhibit odd-even alternation, and a variety of paraffin chain compounds [167] exhibit odd-even alternation in their melting points and heats of fusion, which of course also involves transition from the solid crystalline state to the liquid state. Such alternations have also been exhibited in

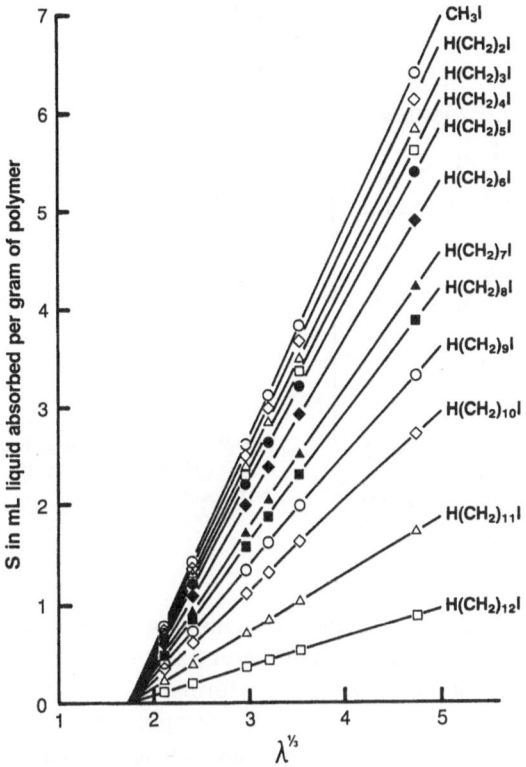

Fig. 27. Correlation of the volume S in mL of sorbed $I(CH_2)_nH$ liquids per gram of polymer with the corresponing cube root of the number (λ) of backbone carbon atoms between crosslink junctions in $(Sty)_{1-x}(DVB)_x$ polymers as a function of n

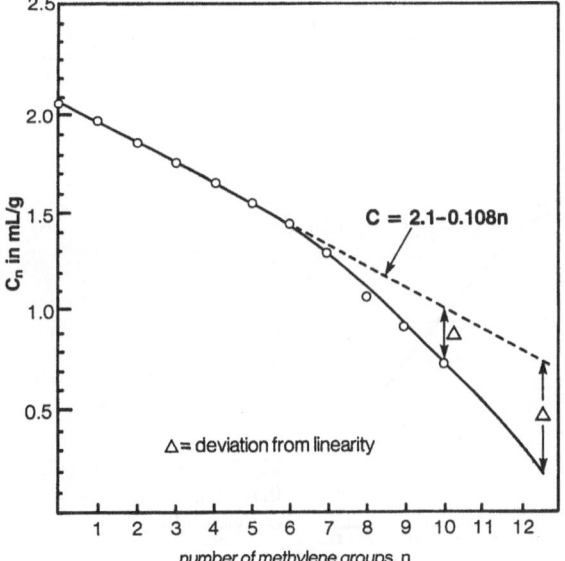

Fig. 28. Correlation of the relative swelling power (C_n) of $Ph(CH_2)_nH$ liquids with the corresponding n

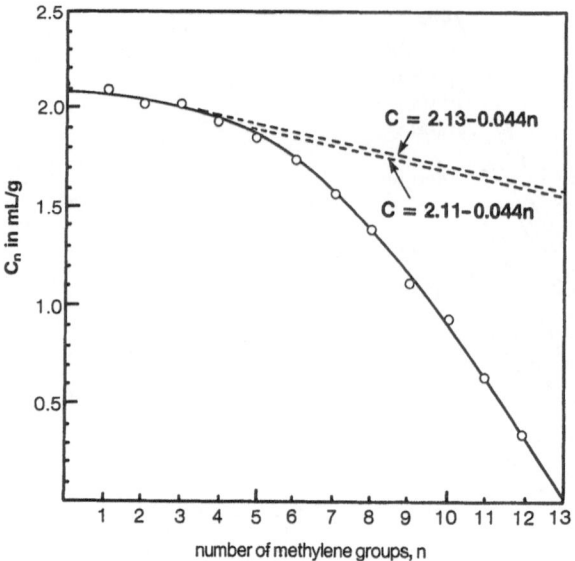

Fig. 29. Correlation of C_n for $I(CH_2)_nH$ liquids with the corresponding n

transition temperatures of some condensed monolayers containining tilted chains [168], and even in the Krafft point and the critical micelle concentrations of sodium alkylsulfonates [169], when the free-energies of micelle formation for these homologous series are taken into consideration. It was concluded by these investigators [170] that there is an ordered orientation in the lamellar structure of micelles generated by alkylsulfonates in aqueous media with respect to the surface of the micelle.

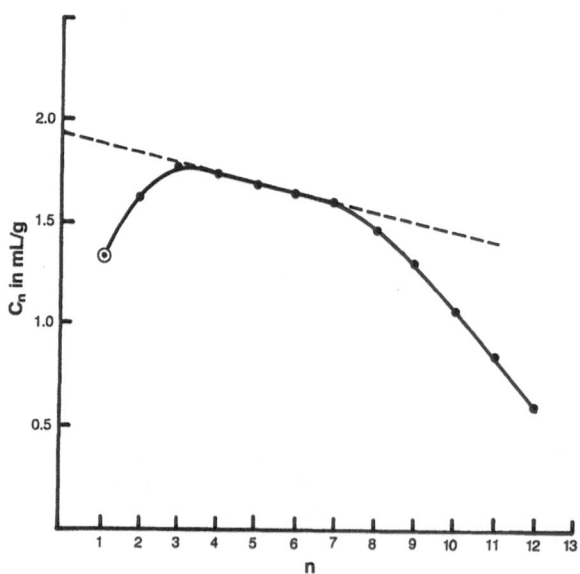

Fig. 30. Correlation of C_n for $Br(CH_2)_nH$ liquids with the corresponding n

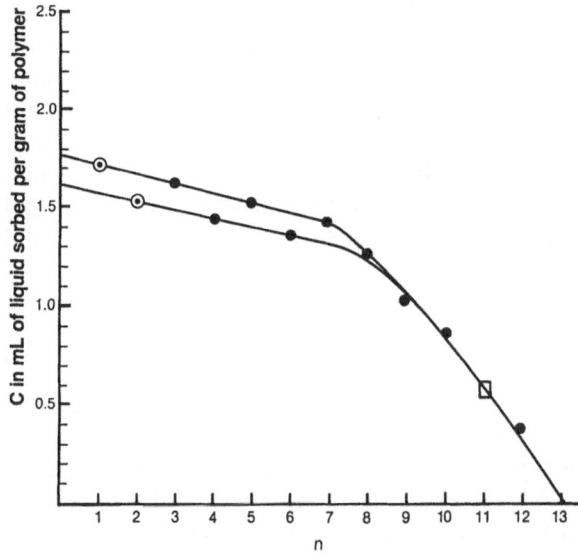

Fig. 31. Correlation of C_n for $Cl(CH_2)_nH$ liquids with the corresponding n

Since ordered structural orientation is believed to be a prerequisite for exhibiting alternation in physical properties in monolayer coverage, I have interpreted the alternation exhibited in the C_n data for the liquids that comprise $Z(CH_2)_nH$ series to mean that such moleclules are fixed to the monomer unit of poly(styrene) segments in the liquid-saturated gel-state via the Z substituent, and that these adsorbed molecules are distributed around the adsorption sites (i.e. the monomer units) in a well-defined orientation with respect to each site, such that the established orientation relative to that monomer unit is maintained despite the freedom of rotation and serpentine movement of the polymer segments between crosslink junctions in the liquid-saturated gel domain.

It would be misleading, however, to interpret how well the molecular structure of the sorbed species is accommodated by the molecular structure of the monomer unit on the basis of the observed swelling power, C, because this parameter is proportional to the product of two variables, namely the number of adsorbed molecules per accessible phenyl group (α), and the molar volume of the sorbed liquid ($V = M/d$), where M and d are the molecular weight and density of the sorbed liquid). Since the density of the liquid reflects how well the molecules in that liquid are accommodated by one another, this effect would be superimposed on the adsorption data. This point of view is consistent with the observations of Fajans [166], who reported that the density of $Z(CH_2)_nH$ liquids exhibit odd-even alternation especially in the lowest three members of a given series. These results amplify the observation made earlier, i.e. that it is more meaningful to interpret on the basis of α how well the molecular structure of the adsorbed species is accommodated by the molecular structure of the monomer unit, rather than on the basis of C from which α is derived (Eq. 15).

The data for α, collected in Tables 5–8, show that α appears to decrease exponentially with n. This is consistent with the point of view that α reflects the dynamic packing density of the sorbed molecules with respect to the adsorption site, namely the styrene units in poly(Sty-*co*-DVB), i.e. it reflects the net result of two opposing forces; molecular attraction and the reactionary force of steric hindrance [from the standpoint of space limitation at the available adsorption site]. The difference in these opposing forces affect the rates of adsorption and desorption exponentially, which should decrease incrementally with n, and therefore Log α_n should be a linear function of n as given by:

$$\text{Log } \alpha_n = \text{Log } \alpha_0 - An \, . \tag{30}$$

The correlations of α_n with n for $Z(CH_2)_nH$ liquids (Figs. 32–35) show that this appears to be the case for those liquids with n < 9. Negative deviation ($\Delta \text{ Log } \alpha_n$) from the linearity expressed by Eq. 30 occurs thereafter, which indicates that yet another factor becomes significant when n becomes greater than 8.

Earlier investigators [171–174], who studied the physical properties of liquids that comprise the homologous series $H(CH_2)_nH$ and related branched alkanes from the standpoint of excess molar volume of mixing [171], light scattering [172], calorimetry [173], and surface tension [174], noted that deviation from the norm established with the first six members of a linear series for a given physical property begins in every case when n becomes greater than 6, and that the magnitude of the deviation from the norm increases with the quantity n − 6. They attributed

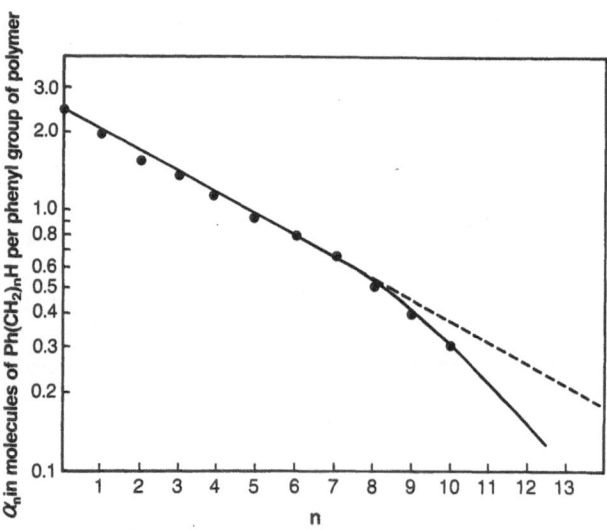

Fig. 32. Correlation of the logarithm of the adsorption parameter, α_n, for $Ph(CH_2)_nH$ liquids with the corresponding n

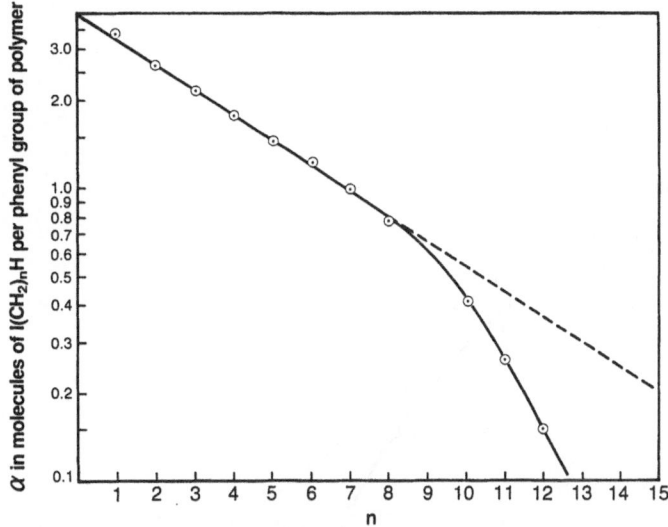

Fig. 33. Correlation of Log α_n for $I(CH_2)_nH$ liquids with the corresponding n

this deviation to self-association of $H(CH_2)_nH$ molecules caused by accumulation of London dispersion forces owing to correlated molecular alignment of the polymethylene segments when n becomes greater than 6. In the case of surface tension studies, Fowkes [174] showed that this cumulative force (F, in 10^{-5} N/cm

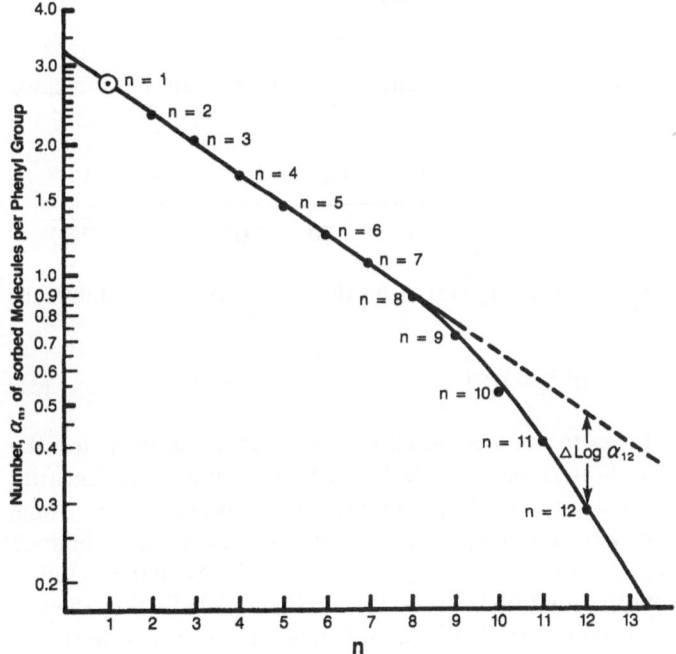

Fig. 34. Correlation of Log α_n for $Br(CH_2)_nH$ liquids with the corresponding n

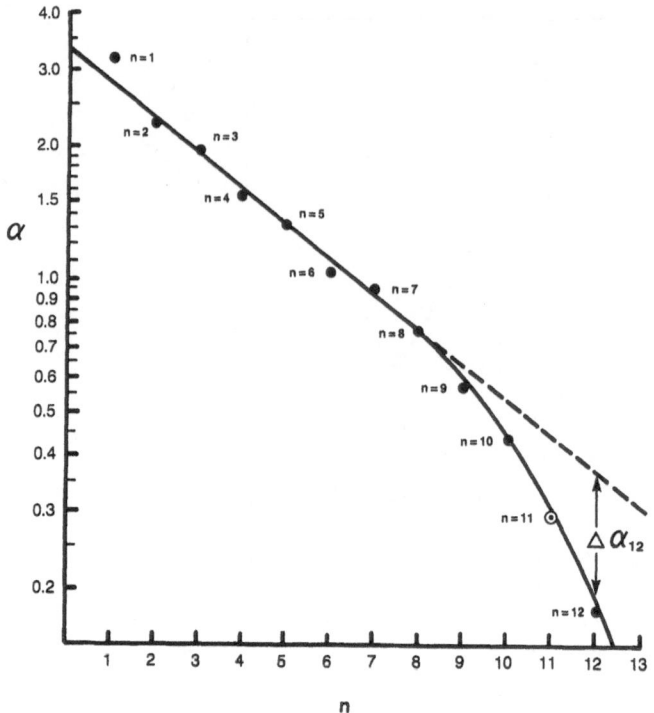

Fig. 35. Correlation of Log α_n for $Cl(CH_2)_nH$ liquids with the corresponding n

or dyn/cm, resp.) of correlated molecular orientation increases with n as indicated below:

n	<7	7	8	9	10	11	12	14	16
F_n	0.0	0.38	0.87	1.16	1.46	1.70	2.04	2.47	2.89

Thus, this set of data for F_n increases linearly with the square root of the difference n − 6 as given approximately by:

$$F_n = 1.15(6 - n)^{1/2} - 0.81 . \tag{31}$$

It is assumed that the same cause-and-effect relationship observed for the $H(CH_2)_nH$ liquids [171–174] is responsible for the deviations from linearity (Figs. 32–35) exhibited by α_n for $Z(CH_2)_nH$ liquids, when n becomes $>n'$ (some number greater than 6). This assumption is supported by the results obtained (Fig. 36) when Δ Log α_n observed for $Z(CH_2)_nH$ liquids with n > 6 in these studies is correlated with F_n reported by Fowkes [174] for the $H(CH_2)_nH$ liquids. In each case Δ Log α_n appears to increase linearly with the difference $(F_n - F')$ as given by:

$$Log \alpha_n = B(F_n - F') , \tag{32}$$

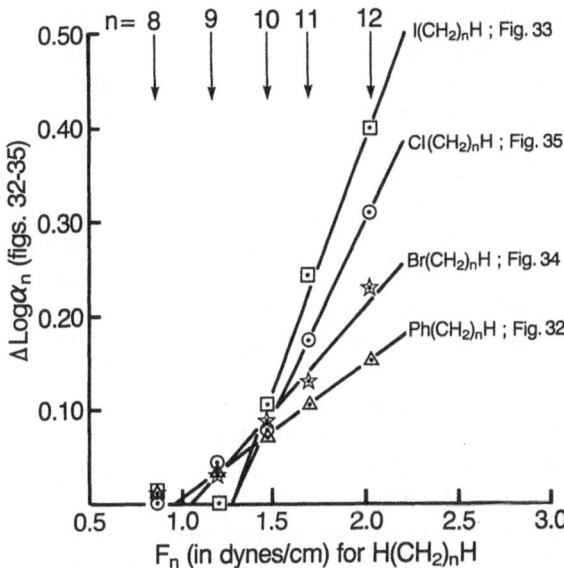

Fig. 36. Correlation of $\Delta \log \alpha_n$ for $Z(CH_2)_nH$ liquids with $n > 6$ (Figs. 32 to 35), with the corresponding force (F_n in dynes/cm) of correlated molecular orientation as measured by Fowkes [174] for $H(CH_2)_nH$ liquids

where F' is the force corresponding to n' where deviation from linearity first becomes significant in a given Z series (i.e. at $n = 8$ in the case that Z is phenyl or bromo, and at $n = 9$ in the case that Z is iodo or chloro).

Thus, Eqs. 30–32 describe how α_n for $Z(CH_2)_nH$ liquids varies with Z and n (from $n = 0$ to 12). The parameters α_0, A and B as defined in Eqs. 30–32 are collected in Table 8, which show that each parameter is affected by the interplay of electronic contribution of the Z substituent and the bulk size of that substituent, i.e. steric hindrance on the basis of volume exclusion with respect to the finite area available at an adsorption site.

Table 8. Adsorption constants α_0, A and B (Eqs. 30–32) for $Z(CH_2)_nH$

Z	α_0	A	B	n'	F'
I	3.9	0.085	0.47	9	1.17
Cl	3.4	0.081	0.41	9	1.25
Br	3.2	0.069	0.23	8	1.07
Ph	2.4	0.083	0.14	8	0.91
H	0	–	–	7	0.38

α_0 is the number of adsorbed $Z(CH_2)_nH$ molecules per phenyl group of polymer at liquid saturation when n is 0.
A is the decrementation constant per added CH_2 group (Eq. 30).
B is the self-association constant [relative to B for $H(CH_2)_nH]$ as defined in Eqs. 30 and 32.
n' is the value of n extrapolated to where $\Delta \log \alpha_n$ first becomes significant as noted in Fig. 36.
F' is the force of correlated molecular orientation (F) at n' (Eqs. 31 and 32).

The constants α_0 (the apparent "inherent" number of adsorbed molecules per accessible phenyl group in the polymer at liquid-saturation when steric hindrance is not a factor, i.e. when $n = 0$) for the $Z(CH_2)_nH$ series as a function of Z are in the order $I > Cl > Br > Ph \gg H$; the constants A (the magnitude of decrementation in Log α_n per added methylene group, reflecting the incremental increase in steric hindrance per added methylene group as defined in Eq. 30) as a function of Z are in the order $I \cong Ph \cong Cl > Br$; the constants B (the magnitude of deviation from linearity as expressed by Eq. 30 per unit change in F_n as defined in Eq. 32) as a function of Z are in the order $H > I > Cl > Br > Ph$; and n' (the number of methylene groups in the polymethylene chain where Δ Log α_n first becomes significant; Eq. 32) as a function of Z are in the order $I \cong Cl > Br \cong Ph > H$.

The above results are consistent with a molecular model of adsorption that assumes liaison of substituent Z of $Z(CH_2)_nH$ sorbates with the phenyl group of poly(Sty-co-DVB) absorbents. When Z is a halogen atom, this liaison is postulated to involve the non-bonded electrons on Z with the pi-electrons of a phenyl group as indicated in Fig. 37a (when the adsorbed molecule is represented by ZCRR'R'') and in Fig. 37b (when ZCRR'R'' is $Z(CH_2)_nH$). The order observed for α_0 (Eq. 30) as a function of Z is consistent with the assumption that the "inherent" dynamic adsorption density of such molecules on the adsorption site (in this case the phenyl groups of the polymer) varies with the polarity and polarizability of substituent Z, and inversely with the "bulkiness" of that substituent.

The decrementation constant A in Eq. 30, i.e. the incremental decrease in Log α_n per additional methylene group exhibited by those $Z(CH_2)_nH$ liquids with $n < 8$, is consistent with the point of view that A reflects the increase in steric hindrance in proportion to the bulk of each additional CH_2 group. When n in the polymethylene chain becomes some number greater than 6, designated n' (i.e. where the cumulative force of correlated molecular orientation first becomes significant), the magnitude of decrementation per additional CH_2 becomes progressively greater than A, because the self-association caused thereby serves

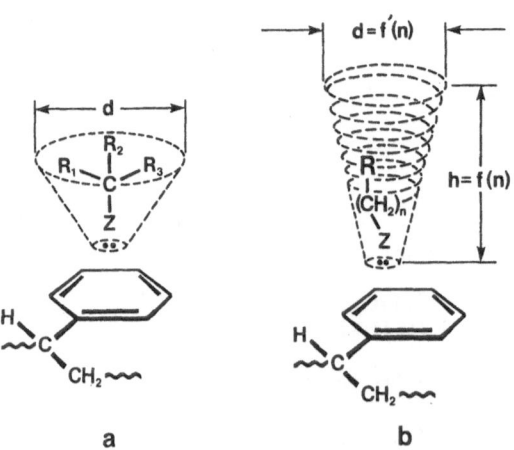

Fig. 37 a, b. Schematic representation of association (**a**) between the pendent phenyl group of poly(styrene) and molecule ZCRR'R'', and (**b**) the corresponding association when ZCRR'R'' is $Z(CH_2)_nH$

to increase the apparent "bulkiness" of the polymethylene chain. This, in effect, increases the effective diameter of the "cone of steric hindrance" generated by prolated gyration above the adsorption site (Fig. 37b), and reduces accordingly Log α_n for those liquids with n > n'. The magnitude of added decrementation (B; Eq. 32), however, appears to vary inversely with the bulkiness of substituent Z, which implies that the ease of self-association owing to correlated molecular orientation of $(CH_2)_nH$ chains with n > 6 is affected by the nature of the substituent Z in the 1-position.

In summary the results observed in these studies of poly(Sty-co-DVB) swelling in $Z(CH_2)_nH$ liquids confirm that the adsorption parameter, $\alpha_{z,n}$, for such liquids is the product of an electronic and a steric factor. The electronic factor, which reflects the affinity of the Z substituent in the liquid for the phenyl group in the polymer, is here the more important one. The steric factor is secondary in that it causes a decrease in $\alpha_{z,n}$ with each additional methylene group added to the omega carbon atom. For liquids with n < 9, in a given Z series, the logarithms of $\alpha_{z,n}$ appear to decrease by an incremental amount, $A_{z,n}$, per methylene group (Eq. 30), which is characteristic of the Z substituent at the alpha carbon atom. For liquids with n > 8 the magnitude of decrease is greater than that expected on the basis of Eq. 30, owing to self-association of $Z(CH_2)_nH$, which becomes significant when n is about 9. The magnitude of deviation from linearity as expressed by Eq. 30 is proportional to the magnitude of self-association caused by correlated molecular orientation of $(CH_2)_{n>6}$, which is dependent upon the characteristics of Z.

3.4.2.2 Non-Linear Alkanes, ZCRR'R''

The adsorption data accumulated for $Z(CH_2)_nH$ liquids with n < 8 (Tables 4–7) were then used as a data-base for comparison with the corresponding adsorption data (Tables 9–11) accumulated subsequently for ZCRR'R'' liquids [175], where R, R' and R'' are either H or a non-cyclic alkyl group, to show how the interplay of steric and electronic factors are affected by systematic modification of CRR'R''. Figures 38–41 compare the correlations of Log $\alpha_{z,N}$ for $Z(CH_2)_nH$ liquids with n < 8 and for $ZCH_{3-m}(CH_3)_m$ liquids with m = 0 to 3 (Table 10) with respect to the corresponding number, N, of carbon atoms in the alkyl group, CRR'R'', attached to substituent Z. These correlations show that the data fall on two straight lines that share a common point at N = 2. The slope of the line of best fit that passes through the data for $ZCH_{3-m}(CH_3)_m$ is in all cases much greater than that which passes through the data for the corresponding $Z(CH_2)_nH$ liquids with n > 1 but <8. The linear relationship for the former and the latter sets of data are expressed respectively by Eqs. 33 and 34:

$$Log\, A_{z,N} = Log\, \alpha_{z,0} - A_{z,m}N, \tag{33}$$

$$Log\, \alpha_{z,N} = Log\, \alpha_{z,2} - A'_{z,n}N, \tag{34}$$

where $A'_{z,n}$ for liquids with n = 2 to 7 in Eq. 34 is less than the corresponding $A_{z,n}$ for liquids with n = 0 to 7 as expressed by Eq. 30. When α_1 is considered

Table 9. Sorption of $ZCH_{(3-m)}(CH_3)_m$ liquids by poly(Sty-*co*-DVB)

Z	m	$C_{z,m}$	$\alpha_{z,m}$	δ_{zm}	χ_1
Ph	0	2.02	1.98	9.16a	0.268
	1	1.84	1.55	8.93a (9.06b)	0.378
	2	1.65	1.23	8.76b	0.494
	3	1.47	0.99	8.57b	0.603
I	0	2.08	3.45	10.1b	0.231
	1	2.03	2.64	9.69b	0.262
	2	1.85	1.91	8.84b	0.372
	3	1.63	1.43	8.57b	0.506
Br	0	[1.34]	[2.72]	10.85b	0.683
	1	1.62	2.26	9.81b	0.512
	2	1.65	1.83	8.40b	0.494
	3	1.49	1.34	8.16b	0.890
Cl	0	[1.70]	[3.21]	10.30b	0.463
	1	[1.54]	[2.23]	9.70b	0.551
	2	1.26	1.44	8.07a (8.04b)	0.731
	3	0.98	0.94	7.85b	0.902

C, α, and χ_1 are as defined in the footnotes of Table 5.
$\delta_{z,m}$ is the solubility parameter (a) reported by Hoy [34] (b) calculated by the method of component contributions in the manner outlined by Van Krevelen [27].

Table 10. Sorption of $ZCRR'(CH_2)_nH$ liquids by poly(Sty-*co*-DVB)

Z	R	R'	n	$C_{z,n}$	$\alpha_{z,n}$	$\delta_{z,n}$	$\chi_{z,n}$
Ph	H	Me	2	1.69	1.13	8.80b	0.469
	Me	Me	2	1.48	0.91	8.62b	0.597
I	H	Me	2	1.85	1.67	8.80b	0.372
Br	H	Me	2	1.65	1.57	8.64a (8.42b)	0.494
	H	Me	3	1.58	1.30	8.63c	0.536
	Me	Me	2	1.56	1.27	8.41c	0.548
	H	Et	2	1.63	1.37	8.63c	0.506
Cl	H	Me	2	1.28	1.25	8.11a (8.12b)	0.719
	Me	Me	2	1.00	0.84	7.95b	0.890
	Me	Et	2	1.05	0.80	8.02b	0.860

C, α, and χ are as defined in the footnotes of Table 5.
$\delta_{z,n}$ is the solubility parameter,
(a) reported by Hoy [34]
(b) calculated by the method of component contributions in the manner outlined by Van Krevelen [27]
(c) calculated as in "b", but beginning with the value reported by Hoy for $BrCH(CH_3)(CH_2)_2H$ instead of the corresponding calculated value, which is assumed to be too low.
R and R' are H, methyl (Me), or ethyl (Et) substituents.

Table 11. Sorption of $Z(CH_2)_nCH(CH_3)_2$ liquids by poly(Sty-*co*-DVB)

Z	n	$C_{z,n}$	$\alpha_{z,n}$	$\delta_{z,n}$	$\chi_{z,n}$
Ph	1	1.62	1.07	9.95a	0.512
I	1	1.83	1.65	10.5a	0.384
Br	1	1.57	1.59	10.1a	0.542
	2	1.46	1.26	9.82a	0.609
Cl	1	1.35	1.34	9.82a	0.677

C, α, and χ are as defined in Table 5.
(a) calculated by the method of component contributions in the manner outlined by Van Krevelen [27].

to be part of the set comprised of the linear molecules as expressed by Eq. 30, the slope calculated by linear regression is correspondingly greater as noted above.

The reason why $A_{z,m}$ is greater than $A'_{z,n}$ is that the former represents the magnitude of decrementation per added methylene group caused by addition of such groups to the carbon atom alpha to the Z substituent, which contributes to both the steric and electronic factors, whereas the latter represents that caused by addition of such groups to the carbon atom omega to the Z substituent, which contributes only to the steric factor. Consequently the difference $(A_{z,m} - A'_{z,n})$ reflects the portion of $A_{z,m}$ attributable to the electronic contribution, and the ratio $(A_{z,m} - A'_{z,n})/A'_{z,n}$ reflects the relative contributions of these two effects on $A_{z,m}$ observed for the corresponding $ZCH_{3-m}(CH_3)_m$ series, which includes the first two members of the $Z(CH_2)_nH$ series.

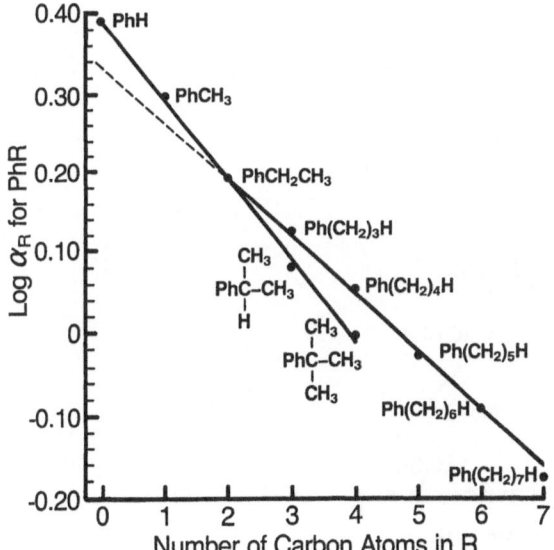

Fig. 38. Correlation of Log α_R for PhR with the number of carbon atoms in R

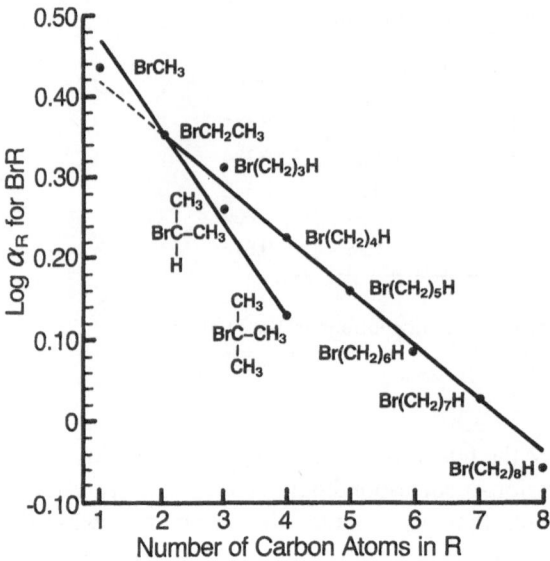

Fig. 39. Correlation of Log α_R for IR with the number of carbon atoms in R

Equations 33 and 34 were then used [175] as base lines for comparison with data accumulated for liquids that represent systematic modification of a member of either $ZCH_{3-m}(CH_3)_m$ or $Z(CH_2)_nH$ by addition of methylene groups to carbon atoms that are in positions ranging from beta to omega from the Z substituent.

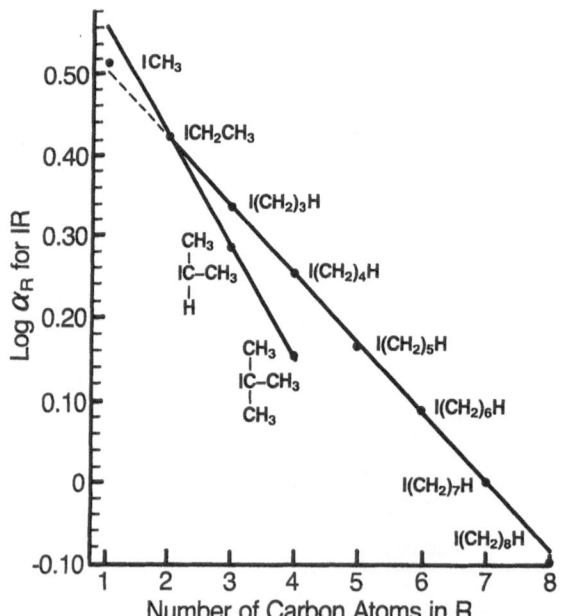

Fig. 40. Correlation of Log α_R for BrR with the number of carbon atoms in R

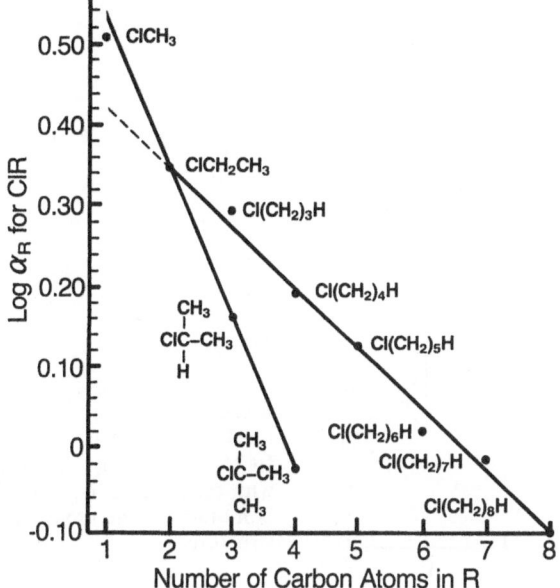

Fig. 41. Correlation of Log α_R for ClR with the number of carbon atoms in R

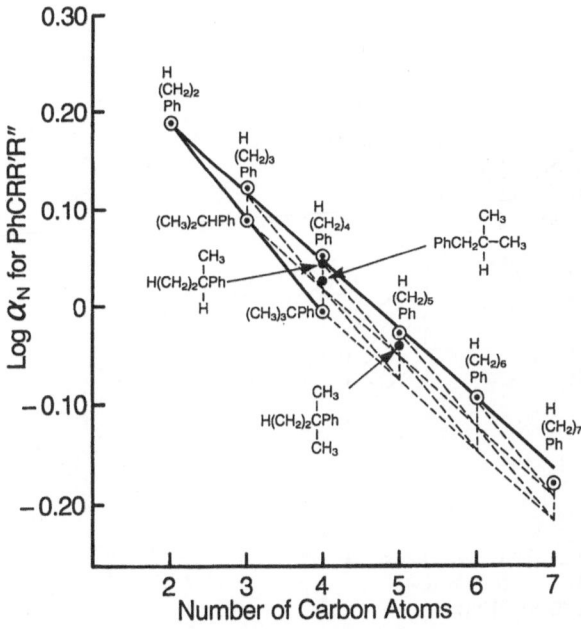

Fig. 42. Correlation of Log $\alpha_{Ph,N}$ for $PhCRR'(CH_2)_nH$ and $Ph(CH_2)_nCH(CH_3)_2$ liquids with the number, N, of carbon atoms in the alkyl group attached to the phenyl substituent

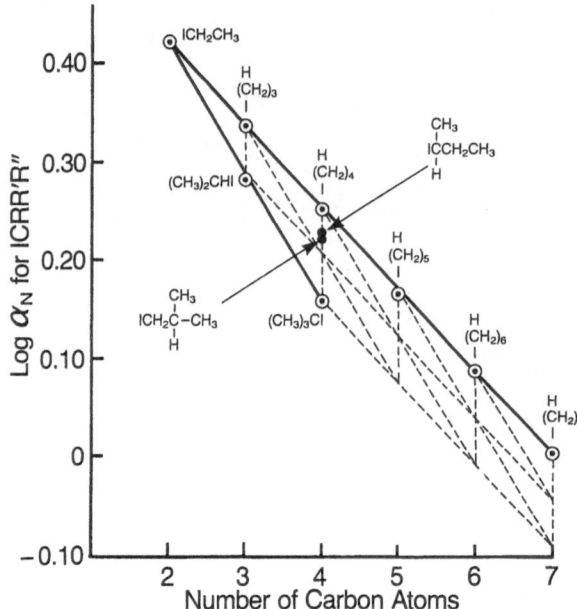

Fig. 43. Correlation of $\alpha_{I,N}$ for ICRR′(CH$_2$)$_n$H and I(CH$_2$)$_n$CH(CH$_3$)$_2$ liquids with the number, N, of carbon atoms in the molecular structure

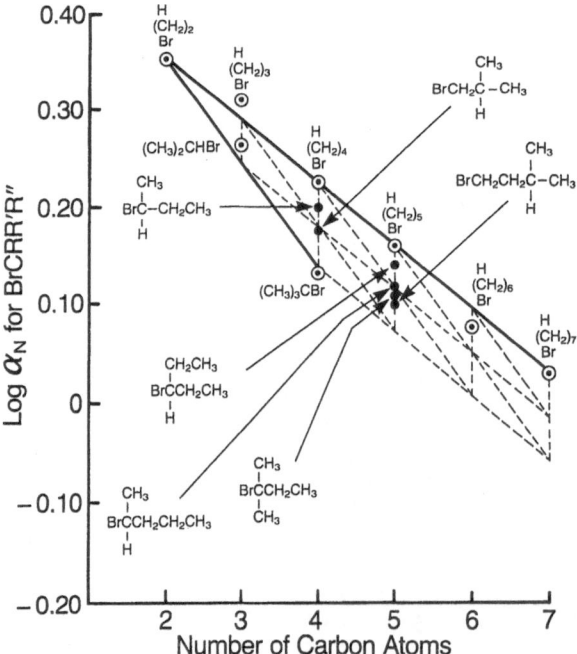

Fig. 44. Correlation of Log $\alpha_{Br,N}$ for BrCRR′(CH$_2$)$_n$H and Br(CH$_2$)$_n$CH(CH$_3$)$_2$ liquids with the number, N, of carbon atoms in the molecular structure

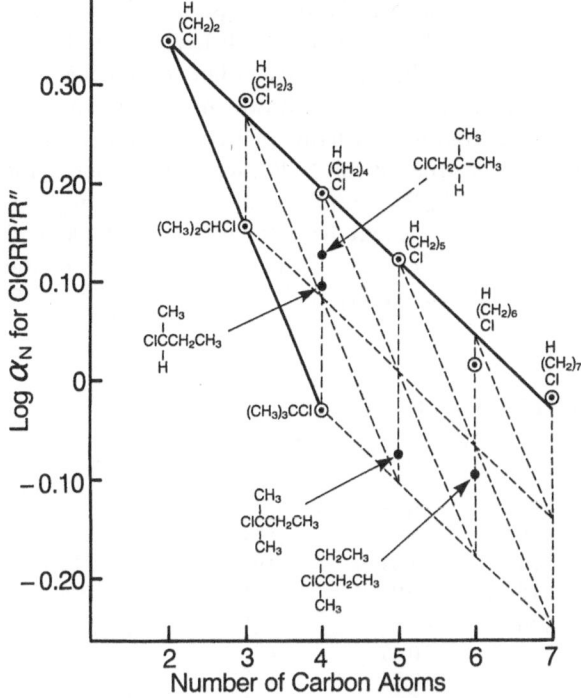

Fig. 45. Correlation of Log $\alpha_{Cl,N}$ for $ClCRR'(CH_2)_nH$ and $Cl(CH_2)_nCH(CH_3)_2$ liquids with the number, N, of carbon atoms in the molecular structure

The adsorption data for the latter liquids are collected in Tables 10 and 11. Figures 42–45 show correlations of Log $\alpha_{z,N}$ for these liquids with the corresponding number, N, of carbon atoms in the alkyl substituent attached to Z. The base lines of reference are represented by the solid lines that pass through the set of data (empty circles) for the two sets of $Z(CH_2)_nH$ and $ZCH_{3-m}(CH_3)_m$ liquids. The dashed lines were drawn parallel to these reference lines to serve as visual-aids for adjudication of parallelism of the data (filled circles) accumulated for the

Table 12. Estimated $\alpha_{z,m,n}$ for $ZCH_{(2-m)}(CH_3)_m(CH_2)_nH$ liquids

Z	m	n = 2	n = 3	n = 4	n = 5
Ph	1	[1.13]	0.95	0.80	0.69
I	1	[1.67]	1.38	1.16	0.94
Br	1	[1.57]	[1.30]	1.12	0.98
Cl	1	[1.25]	1.04	0.87	0.73
Ph	2	[0.91]	0.75	0.63	0.54
I	2	1.25	1.05	0.85	0.68
Br	2	[1.27]	1.10	0.94	0.82
Cl	2	[0.84]	0.72	0.60	0.51

[ ] data from Table 10 used as starting point for estimation of the data recorded here for the given Z, m series.

test-liquids listed in Tables 10 and 11. These correlations (Figs. 42–45) show that the magnitude of depression from the reference line established by $Z(CH_2)_nH$ liquids, and the displacement to the right of reference line established for $ZCH_{3-m}(CH_3)_m$ liquids is in all cases commensurate with the magnitude of steric hindrance expected on the basis of the molecular structure of the test-liquid [175].

If one now assumes that this will be true for all $ZCRR'R''$ liquids, then α for $ZCH_{2-m}R_m(CH_2)_nH$ and $Z(CH_2)_nCH_{3-m}R_m$ liquids that have not yet been established by experimental measurement might be estimable on the basis of expected parallelisms [175] as noted in Figs. 42–45. As examples some of the α-values estimated for liquids in the above two types of homologous series [175] are collected in Tables 12 and 13.

Table 13. Estimated $\alpha_{z,m,n}$ for $Z(CH_2)_nCH_{(3-m)}(CH_3)_m$ liquids

Z	m	n = 1	n = 2	n = 3	n = 4
Ph	2	[1.07]	0.91	0.76	0.64
I	2	[1.65]	1.38	1.15	0.94
Br	2	[1.50]	[1.26]	1.07	0.91
Cl	2	[1.34]	1.12	0.95	0.81
Ph	3	0.84	0.70	0.61	0.51
I	3	1.23	1.02	0.85	0.71
Br	3	1.17	1.01	0.87	0.75
Cl	3	0.85	0.71	0.61	0.51

[ ] data from Table 11 used as a starting point for estimation of the $\alpha_{z,m,n}$ data recorded here for the given Z, m series.

3.4.3 Mono-Substituted Cyclic Alkanes, $ZCR(CH_2)_n$

The adsorption data accumulated for these liquids [175] are collected in Table 14. If one assumes that steric hindrance is the only factor that needs to be considered in

Table 14. Sorption of $ZR(CH_2)_n$ liquids by poly(Sty-CO-DVB)

Z	R	n	$C_{Z,R,n}$	$\alpha_{Z,R,n}$	$\delta_{Z,R,n}$	x_1
Ph	H	5	1.80	1.11	9.75	0.402
I	H	5	1.78	1.43	10.08	0.414
Br	H	4	1.97	1.91	9.95	0.298
	H	5	1.92	1.62	9.76	0.329
	H	6	1.96	1.49	9.61	0.304
	CH_3	5	1.83	1.37	9.39	0.394
Cl	H	4	1.95	1.95	9.74	0.311
	H	5	1.89	1.66	9.57	0.347

C, α, and x_1 are as defined in footnotes of Table 5.
$\delta_{Z,R,n}$ was calculated by the method of component contributions in the manner outlined by Van Krevelen [27].

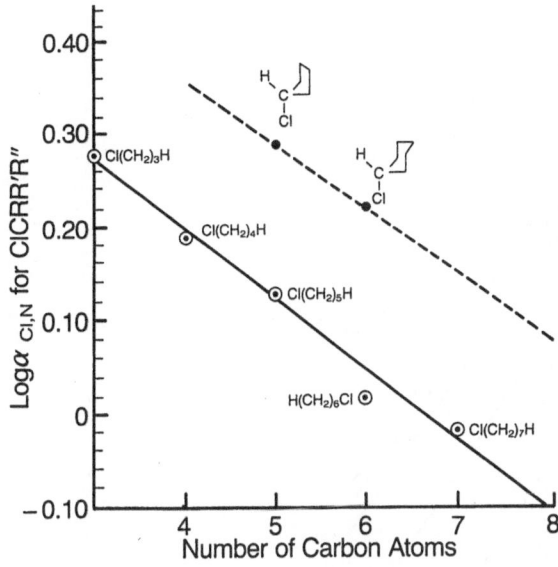

Fig. 46. Correlation of Log $\alpha_{Cl,N}$ for cyclic $ClCH(CH_2)_n$ liquids with the number, N, of carbon atoms in the molecular structure

comparing $\alpha_{z,N}$ for a given $ZCH(CH_2)_n$ with the corresponding $ZCH_2(CH_2)_nH$ liquid, then it is reasonable to expect that $\alpha_{z,N}$ for the former set will be between that for latter set and that for $ZCHCH_3(CH_2)_{n-2}H$.

The correlations of Log $\alpha_{z,N}$ for the cyclic series and for the corresponding acyclic series with N (Figs. 46–48) show that this is not the case. The set of $\alpha_{z,N}$ for the former are in each case uniformly much greater that those in the

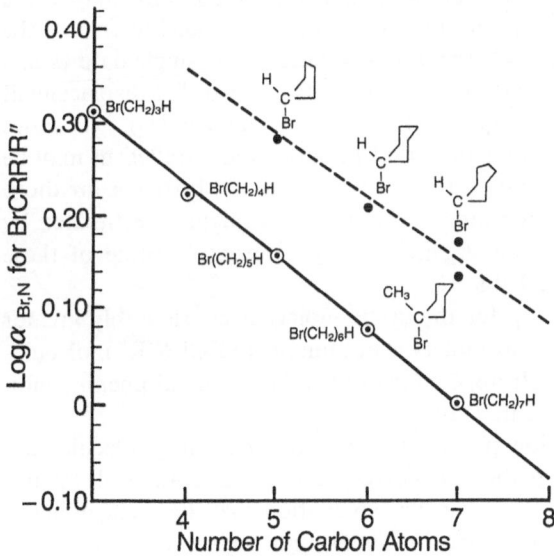

Fig. 47. Correlation of Log $\alpha_{Br,N}$ for cyclic $BrCR(CH_2)_n$ liquids with the number, N, of carbon atoms in the molecular structure

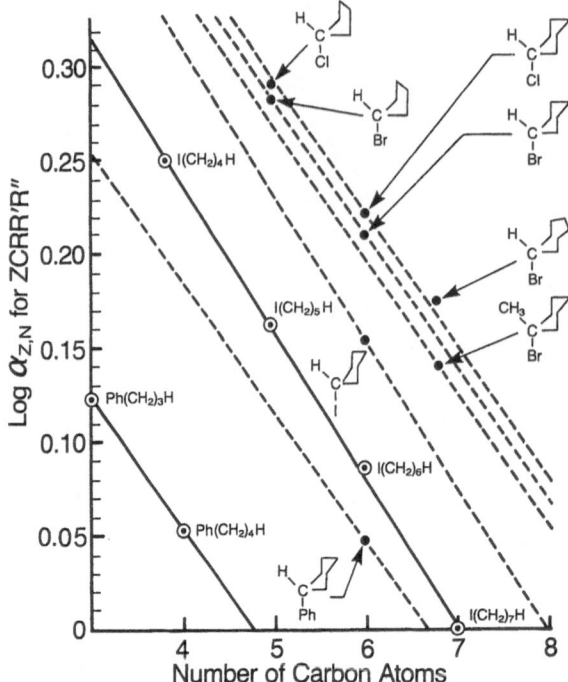

Fig. 48. Comparison of Log $\alpha_{z,N}$ for cyclic $ZCR(CH_2)_n$ liquids with the corresponding data for $ZCHR(CH_2)H$ liquids

corresponding set of the acyclic series. In the cases for which two or more $\alpha_{z,N}$ values for a given $ZCR(CH_2)_n$ series were determined, i.e. for $Z = Br$ and Cl, the data for the cyclic liquids fall on a line that appears to be parallel to that of the corresponding acyclic series (Figs. 46 and 47). [Presumably the single data points established for $ICH(CH_2)_5$ and $PhCH(CH_2)_5$ identify the parallel displacement from the respective linear relationships for the corresponding $Z(CH_2)_{n+1}H$ series as shown in Fig. 48.] It appears, therefore, that the observed parallelism may be general for cyclic $ZCR(CH_2)_n$ liquids with n between 3 and 8. If true, α for those liquids that have not yet been determined experimentally might be estimated by extrapolation [175] as noted above. Again by way of example some of these estimable values are collected in Table 15.

The above observation that $\alpha_{z,N}$ for the cyclic moiety is considerably greater than that of the corresponding linear moiety is not unique to ZCRR′R″ molecules in which Z is phenyl of halogen. It appears instead to be a general phenomenon as indicated by the data collected in Table 16.

The reason why the adsorption parameters, α_c, for the cyclic molecules are significantly greater than those, α_L, for the corresponding linear molecules is not yet clear [175]. There are at least four possible explanations: (1) the cyclic moieties can pack more efficiently in a monolayer on or around a given adsorption site

Table 15. Estimated $\alpha_{Z,R,n}$ for $ZCR(CH_2)_n$ liquids

Z	R	n = 3	n = 4	n = 5	n = 6	n = 7
Ph	H	1.53	1.30	[1.11]	0.95	0.79
I	H	2.03	1.70	[1.43]	1.18	0.98
Br	H	2.26	[1.91]	[1.62]	[1.49]	1.26
	CH$_3$	2.19	1.57	[1.37]	1.14	0.97
Cl	H	2.32	[1.95]	[1.66]	1.46	1.28

[ ] data taken from Table 14 and used as a starting point for estimation of the $\alpha_{z,n}$ data recorded here for the given Z, n series.

Table 16. Comparison of α_c for cyclic molecules with α_L for a corresponding linear molecule

Cyclic	α_c	Linear	α_L	$\alpha_c - \alpha_L$
cyclohexanol	0.47	$H(CH_2)_6OH$	<0.01	0.46
cyclohexane	0.56	$H(CH_2)_6H$	<0.1	0.55
cyclohexane	1.50	$H(CH_2)_4CH=CH_2$	0.31	1.19
cis-decalin	0.90	$H(CH_2)_{10}H$	<0.01	0.89
cyclopentanone	2.29	$(CH_3CH_2)_2CO$	1.50	0.79
tetrahydrofuran	2.36	$(CH_3CH_2)_2O$	0.64	1.22
phenylcyclohexane	1.11	$Ph(CH_2)_6H$	0.80	0.31
tetrahydronaphthalene	1.69	$Ph(CH_2)_4H$	1.13	0.56
iodocyclohexane	1.43	$I(CH_2)_6H$	1.22	0.21
bromocyclohexane	1.62	$Br(CH_2)_6H$	1.22	0.40
chlorocyclohexane	1.66	$Cl(CH_2)_6H$	1.04	0.62

[represented in these studies by the phenyl group of poly(sty-co-DVB) at liquid saturation], (2) the cyclic moieties can pack in multiple layers on these adsorption sites, (3) the cyclic moieties associate strongly, not only with the phenyl group of the monomer unit, but also with the rest of the molecular structure of that unit; i.e. the area of the adsorption is increased from that offered by a phenyl group to that offered by the entire monomer unit, and (4) these cyclic moities may have a much higher polarizability than expected (perhaps due to their higher density). Obviously more study is needed to determine which if any of the four possibilities is correct, and to elucidate "why" and "how" adsorption of the cyclic molecules is different from the mode of adsorption for the corresponding non-cyclic molecules.

4 Correlations of α and C with Other Parameters

Section 2 reviewed the relevant contributions to the understanding of polymer swelling and permeation reported by earlier investigators. It also discusses the important parameters that have bearing on these phenomena, namely the Hildebrand Solubility Parameters, δ, the Flory-Huggins Interaction Parameters, χ and the Guenet Gel-Parameters, ᾱ. Section 3 reviewed the work carried out in 3M Laboratories that led to the concept of an adsorption parameter (α in molecules per monomer unit of polymer), which is calculated from the observed relative swelling power (C in ml of adsorbed liquid per gram of polymer) for the corresponding liquid with respect to the given polymer as shown in Eqs. 14 and 15. The discussions that follow consider how α and C correlate with the parameters δ, χ, and ᾱ reported by the earlier investigators and show how these parameters can be calculated directly from C or α for the corresponding polymer-liquid system. Actually Guenet's investigation of gel-formation from polymer solutions (1980 to the present) was contemporary with the 3M studies (1978 to the present), but unfortunately for both laboratories they were being conducted without knowledge of the work going on in the other until the beginning of 1989. This of course precluded cross-fertilization of ideas, which would have accelerated progress in both laboratories.

4.1 Hildebrand Solubility Parameter, δ

Hildebrand [21] defined the solubility parameter, δ, of a given liquid to be the square root of the internal energy density, i.e. E in cal/mol, divided by its molar volume, V in mL/mole, as described in Section 1.2.1 of this review, and he noted that the mutual compatibility of two molecular species is maximal when the difference in their respective solubility parameters is zero (provided that entropic effects can be ignored). Subsequent investigators [27–40, 176] have added modifications to his observation in the hope of developing a general relationship that would enable one to calculate mutual compatibility in all cases on the basis of the molecular structure of the species in question. To date this approach has given results that are tenuous at best, but even the limited successes serve to attract continued interest in this parameter, owing to the absence of a suitable alternative that might achieve the much sought after generality.

Since solubility and swellability of a polymer in a given liquid are related phenomena [21, 41], one would expect that the relative swelling power, C in volume of sorbed liquid per gram of polymer, should relate meaningfully to the solubility parameter. This is indeed the case [177, 178] as indicated by the correlation of δ, in $(cal/mL)^{1/2}$, reported by various investigators [28, 34 176] for liquid substituted-benzenes, with the corresponding C (Fig. 49), which shows the expected parabolic relationship. The solubility parameter for polystyrene indicated by the apex of this parabolic relationship is 9.5, where C is maximal at 2.25. This value is in agreement with the value (9.4) reported by Hoy [32], but not with the values reported by others [37–39], which range from 8.6 to 9.7. Similar correlations [177, 178] of C with δ_{liq} reported for aliphatic liquids (Table 17), however, indicate that maximal C occurs at $\delta_{liq} = 9.2$ for multichloro-substituted hydrocarbons, at 9.1 for ketones, at about 8.4 for esters, and at about 7.3 for ethers (Fig. 50). Thus the range of δ_{pol} observed for polystyrene in these five classes of liquids spans the range reported by earlier investigators [27–39], 176, who used mixed classifications of liquids to establish maximal solubility. It was concluded [177–178],

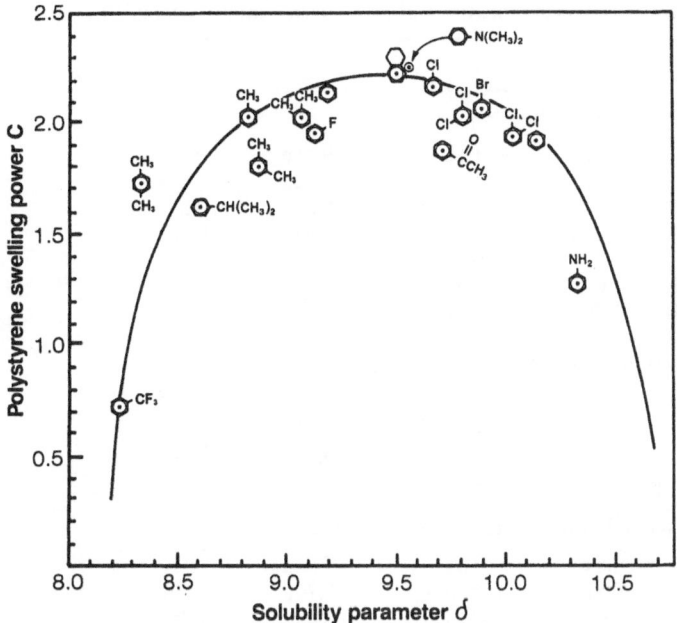

Fig. 49. Correlation of observed relative swelling power, C, of aromatic liquids for poly(Sty-*co*-DVB) with the solubility parameter, δ, reported by Hoy [34] for the corresponding liquid

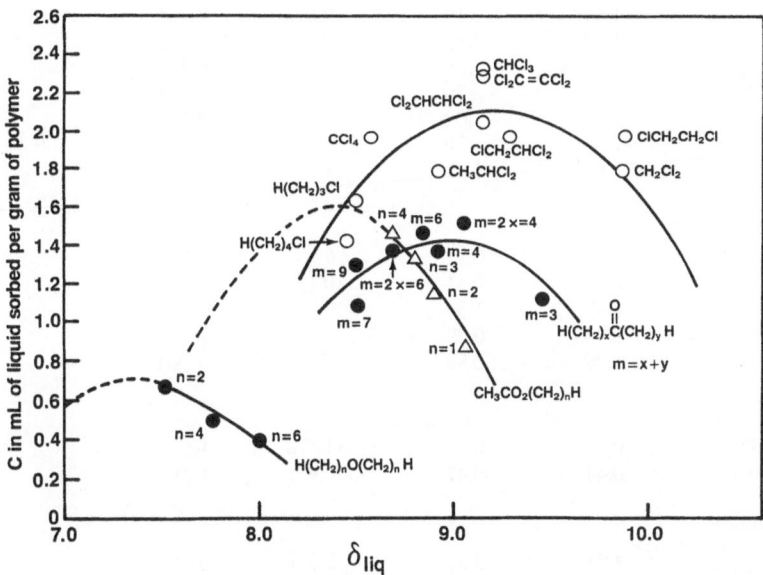

Fig. 50. Correlation of observed relative swelling powers, C, of ethers, esters, ketones, and chlorocarbons for poly(Sty-*co*-DVB) with the solubility parameter, δ, reported by Hoy [34] or by Hansen [176] for the corresponding liquid

Table 17. Parameters for various classifications of polystyrene-liquid systems

Liquid	C	α	δ	χ_1
Ketones: $\delta_{pol} = 9.1$ at maximal C (Fig. 50)				
$CH_3CO(CH_2)_2H$	1.11	1.28	9.45a	0.823
$CH_3CO(CH_2)_3H$	1.37	1.35	8.92a	0.664
$H(CH_2)_2CO(CH_2)_2H$	1.52	1.50	9.06a	0.573
$CH_3CO(CH_2)_5H$	1.47	1.09	8.84b	0.603
$CH_3CO(CH_2)_6H$	1.31	0.87	8.50b	0.701
$H(CH_2)_2CO(CH_2)_7H$	1.04	0.57	8.52a	0.866
$H(CH_2)_3CO(CH_2)_3H$	1.36	1.01	8.70a	0.670
Esters: $\delta_{pol} = 8.4$ at maximal C (estimated graphically in Fig. 50)				
$CH_3CO_2CH_3$	0.86	1.13	9.43a	0.975
$CH_3CO_2(CH_2)_2H$	1.13	1.20	8.91a	0.81
$CH_3CO_2(CH_2)_3H$	1.33	1.20	8.80a	0.689
$CH_3CO_2(CH_2)_4H$	1.46	1.16	8.69a	0.609
Ethers: $\delta_{pol} = 7.3$ at maximal C (estimated graphically in Fig. 50)				
$H(CH_2)_2O(CH_2)_2H$	0.64	0.64	7.53a	1.11
$H(CH_2)_4O(CH_2)_4H$	0.51	0.30	7.76a	1.19
$H(CH_2)_6O(CH_2)_6H$	0.41	0.18	8.01a	1.25
Tetrahydrofuran	2.00	2.57	9.05a	0.28
Multisubstituted Halocarbons: $\delta_{pol} = 9.2$ at maximal C (Fig. 50).				
CH_2Cl_2	1.99	3.27	9.88a	0.286
$CHCl_3$	2.32	2.99	9.16a	0.085
CCl_4	1.97	2.14	8.55a	0.298
CH_3CHCl_2	1.68	2.10	8.92a	0.475
$ClCH_2CH_2Cl$	1.79	2.39	9.86a	0.408
$ClCH_2CHCl_2$	2.04	2.28	9.16a	0.256
$Cl_2CHCHCl_2$	2.29	2.28	9.16a	0.103
$Cl_2C=CCl_2$	1.96	2.02	9.28a	0.304
$Cl(CH_2)_4Cl$	1.78	1.70		0.414
$Br(CH_2)_4Br$	1.82	1.58		0.390
Aliphatic Hydrocarbons				
$H(CH_2)_nH$	0	0	7–8	1.5
Cyclohexane	0.53	0.51	10.42a	1.18
cyclohexene	1.47	1.50		0.60
trans-decalin	0.44	0.29		1.23
cis-decalin	1.00	0.897		0.890
Alcohols				
$H(CH_2)_nOH$	0	0	11–10	1.5
Cyclohexanol	0.47	0.47	10.92a	1.21
Sulfur Compounds				
CS	1.95	3.37	9.92a	0.31

C, α, and χ_1 are as defined in the footnotes of Table 5.
δ is the solubility parameter (a) reported by Hoy [34], and (b) reported by Hansen [176].

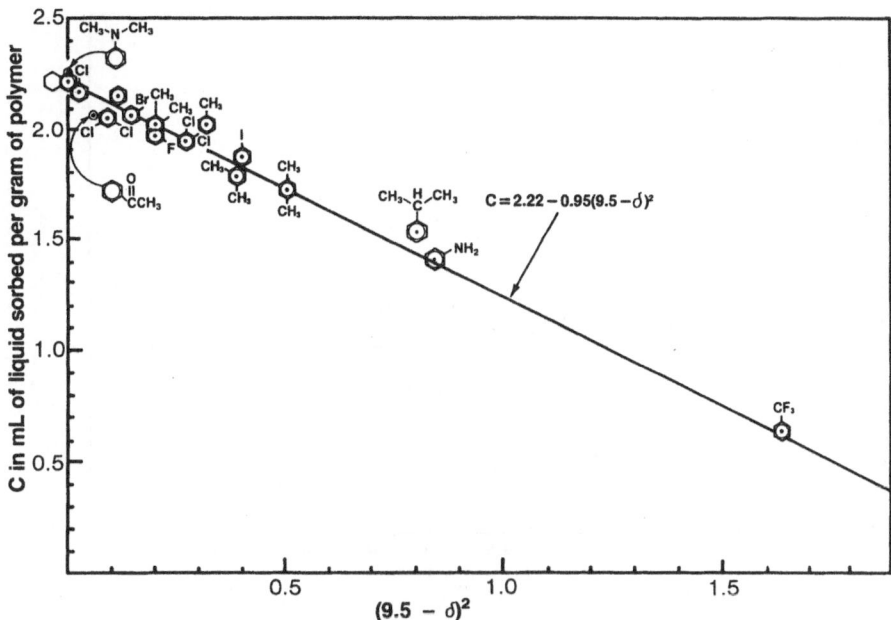

Fig. 51. Correlation of observed relative swelling power, C, for poly(Sty-*co*-DVB) with the corresponding square of the difference (9.5-δ_{liq}), where 9.5 is the observed solubility parameter (δ_{pol}) of the polymer with respect to the set of aromatic liquids shown in Fig. 49; i.e. $\delta_{pol} = \delta_{liq}$ at maximal C

therefore, that the δ_{pol}-data reported by the earlier investigators reflects the arbitrary choice of test-liquids as much as it does the character of polystyrene.

The correlation of C with $(\delta_{pol} - \delta_{liq})^2$ for the set of aromatic liquids ($\delta_{pol} = 9.5$; Fig. 49; Table 2) is a straight line (Fig. 51) given by;

$$C = 2.22 - 0.95(9.5 - \delta_{liq})^2 . \tag{35}$$

Similarly, the correlations of C with $(\delta_{pol} - \delta_{liq})^2$ for the sets of Ph(CH$_2$)$_n$H liquids [160] and the liquid ketones, esters and ethers [177, 178] produce a set of four essentially parallel lines (Fig. 52) given approximately by:

$$C = C_0 - 0.60(\delta_{pol} - \delta_{liq})^2 \tag{36}$$

where the constant C_0 is 2.19 for Ph(CH$_2$)$_n$H, 1.48 for the esters, 1.42 for the ketones, and 0.64 for the ethers, and the respective δ_{pol} are as noted above. These results are consistent with the point of view that C_0 as well as the corresponding observed δ_{pol}, reflects how well the functional group Z in the sorbed liquid ZR is accommodated by the molecular structure of the monomer unit of the polymer, and that C varies inversely with the difference in cohesive energy density, which reflects the molecular structures of the monomer unit and sorbed liquid.

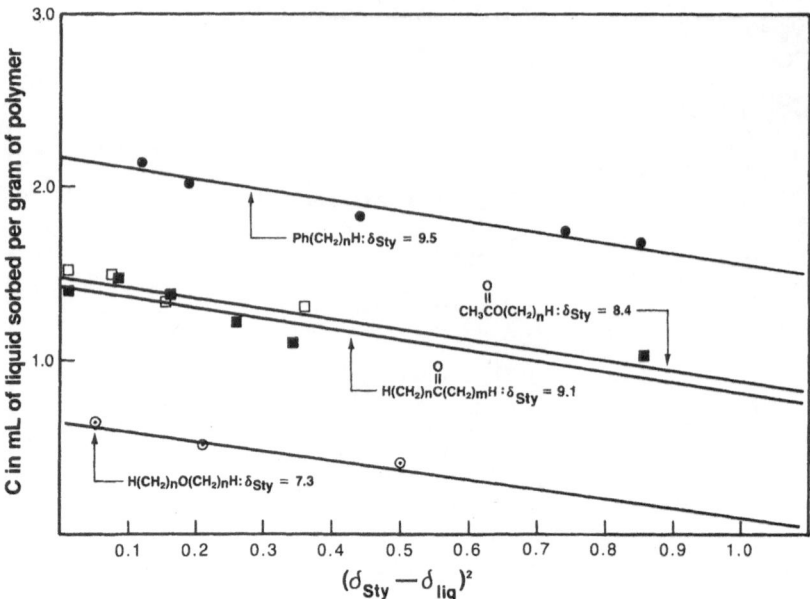

Fig. 52. Correlation of C with the corresponding $(\delta_{pol} - \delta_{liq})^2$, where δ_{pol} is the observed solubility parameter of the polymer with respect to the class of liquids specified in Fig. 50; i.e. 9.5 for substituted benzenes, 8.4 for aliphatic esters, 9.1 for aliphatic ketones, and 7.3 for aliphatic ethers

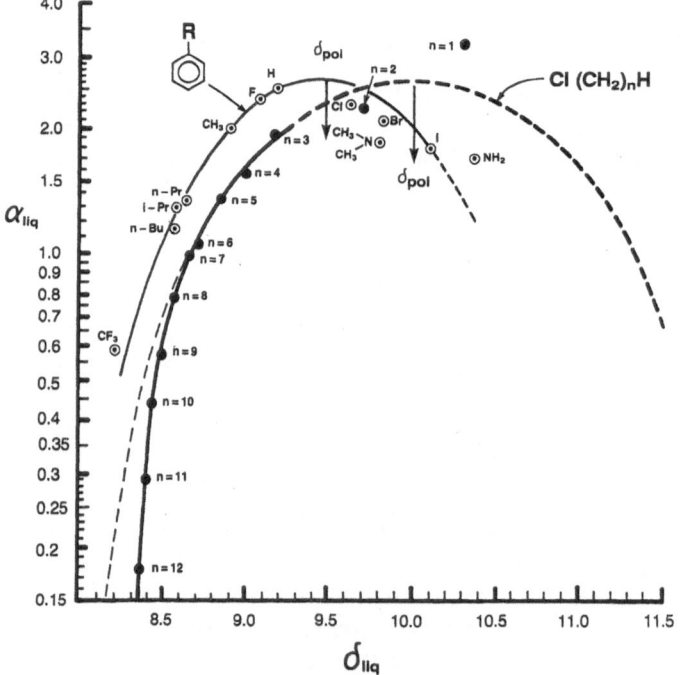

Fig. 53. Correlation of Log α_{liq} with the solubility parameter, δ_{liq}, reported for monosubstituted benzenes, PhR, and calculated for $Cl(CH_2)_nH$ liquids

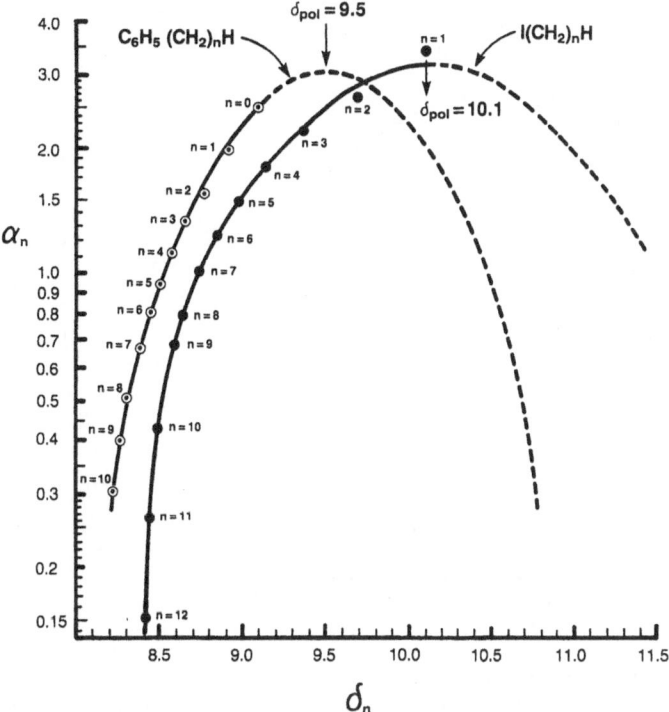

Fig. 54. Correlation of Log α_n with δ_n calculated for the corresponding $Ph(CH_2)_nH$ and $I(CH_2)_nH$ liquids

The molecular nature of such associations is understood better when α_n instead of C_n is correlated with the corresponding difference in cohesive energy density. This is especially true in the case of $Z(CH_2)_nH$ liquids [160–164], despite that the identification of δ_{pol} is relatively imprecise simply because such correlations generate only about half of the expected parabolic relationship, thus not identifying exactly where δ_n is maximal (Figs. 53–55). Since δ_n varies inversely with n in such series, and δ_{max} is equal to or greater than δ_1, δ_{pol} can only be estimated on the basis of symmetry in the expected parabolic relationship. The most probable "correct" δ_{max} for a given set of such liquids is then adjudicated more precisely by successive iterative approximations of δ_{pol} on the basis of the best fit of the data generated thereby to the straight line expressed by:

$$\text{Log } \alpha_n = \text{Log } \alpha_{max} - D(\delta_{pol} - \delta_n)^2 \tag{37}$$

where α_{max} is the maximal value that identifies $\delta_{pol} = \delta_n$ at maximal Log α_n for the given $Z(CH_2)_nH$ series, and D in mL/cal is the decrementation constant per unit difference in cohesive energy density. The constants α_{max}, D and δ_{pol} for each $Z(CH_2)_nH$ series studied thus far are collected in Table 18.

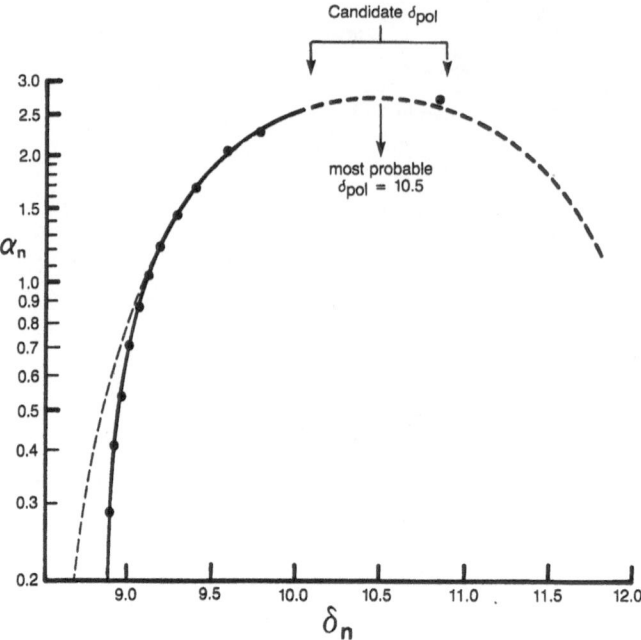

Fig. 55. Correlation of $\log \alpha_n$ with the solubility parameters, δ_n, calculated for the corresponding $Br(CH_2)_nH$ liquids, showing the most probable solubility parameter for the polymer, δ_{pol}, within the range of possible δ_{pol}

Table 18. Effective solubility parameters (δ'_n) for $Z(CH_2)_nH$ liquids with $n > 6$

Z	α_{max}	D	δ_{pol}	δ'_7	δ'_8	δ'_9	δ'_{10}	δ'_{11}	δ'_{12}
Ph (a)	2.7	0.484	9.5	8.32	8.17	8.09	7.98	—	—
Cl (b)	2.8	0.260	10.0	—	—	8.35	8.23	8.03	7.86
Br (c)	3.0	0.233	10.5	9.12	8.99	8.86	8.71	8.57	8.40
I (d)	3.5	0.289	10.1	8.75	8.53	8.33	8.15	7.96	7.83

α_{max}, D, and δ_{pol} are constants as defined in Eq. 37.
(a) δ'_n calculated from Fig. 57 using the data collected in Table 4.
(b) δ'_n calculated from Fig. 56 using the data collected in Table 7.
(c) δ'_n calculated from Fig. 58 using the data collected in Table 6.
(d) δ'_n calculated from Fig. 57 using the data collected in Table 5.

Since the reported δ_n for a given $Z(CH_2)_nH$ series usually represent less than four of the first six members of that series, the rest of the needed δ_n data for the first 12 members were calculated [161–164, 177, 178] by the method of additive contribution of molecular components to the cohesive energy density of the sorbed liquid [27] as noted in Tables 4–7. Despite the uncertainty in identifying δ_{pol}

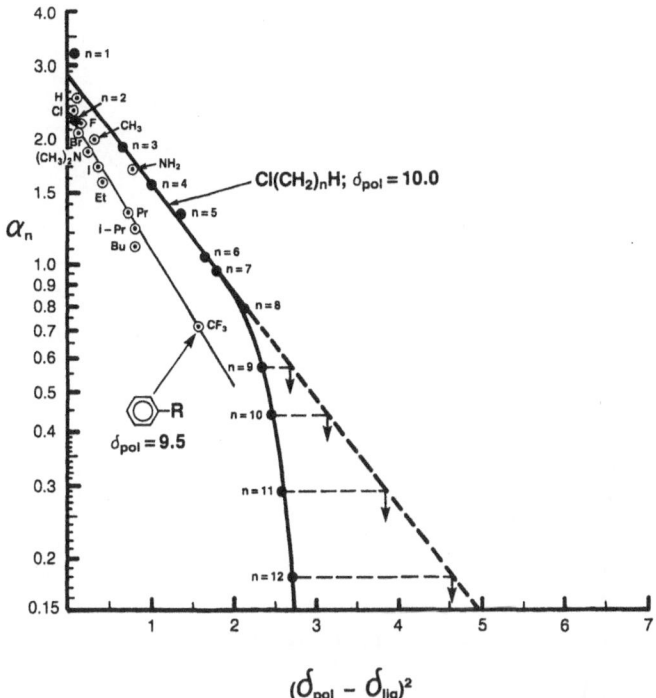

Fig. 56. Correlation of Log δ_{liq} with $(\delta_{pol} - \delta_{liq})^2$ for $Cl(CH_2)_nH$ and PhR liquids

precisely, the correlations (Figs. 56–58) of Log α_n with $(\delta_{pol} - \delta_n)^2$ show that deviation from the linear relationship (Eq. 37), established using only the first six members of a given homologous series, begins when n becomes n' (some number greater than 5 but less than 9), and that the magnitude of this deviation increases with the difference n − 6. These results parallel those observed when Log α_n is correlated with n (Figs. 32–35), which is not surprising since δ_n for $Z(CH_2)_nH$ liquids varies inversely with n. It is suspected, therefore, that self-association, owing to correlation of London dispersion forces that first become significant when n becomes >6, is responsible for deviation from the linearity expressed by Eq. 37 as well as the deviation from linearity expressed by Eq. 30.

These results emphasize that the well-known method of additive contributions of molecular components to cohesive energy [27, 28], which uses component data established for small molecules (i.e. entropic effects not significant) cannot be used reliably to calculate δ for much larger molecules (entropic effects owing to self-association in the liquid state can be quite significant). Thus, deviation from linearity as expressed by Eq. 37 occurs as expected in the $Z(CH_2)_nH$ series of liquids, when these entropic effects become too great to be ignored (in these cases at n', some number greater than 6). It is implied from these observations that any attempt to calculate δ_{pol} by a method that employs such component contributions [36, 176, 179] will lead to conclusions that are at best highly suspect, unless the

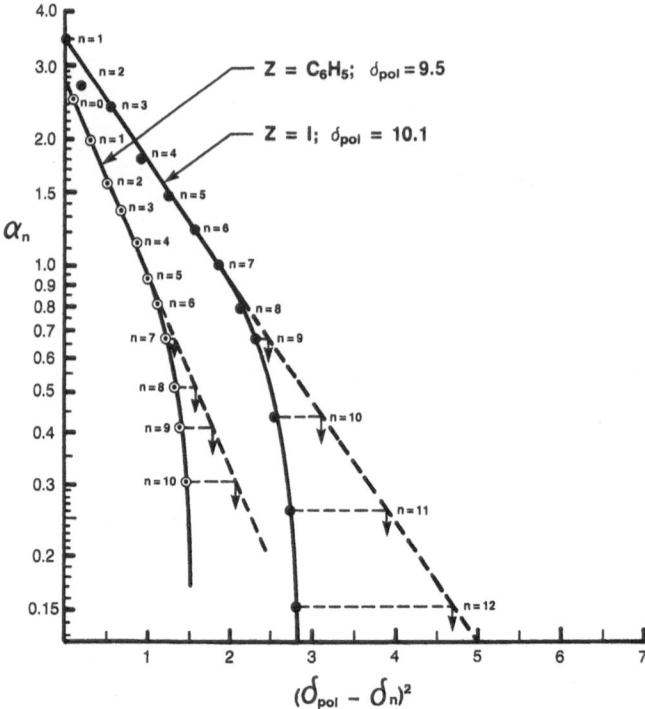

Fig. 57. Correlation of Log α_n with $(\delta_{pol} - \delta_n)^2$ for $I(CH_2)_nH$ and $Ph(CH_2)_nH$ liquids

large contribution of molecular association of polymer with solvent and of polymer with polymer is considered adequately. This caveat is merely an obvious extension of that given early on by Hildebrand [21], who pointed out that his solubility parameter concept applies only in those cases where entropy does not change significantly.

If for calculation purposes one were to assume that the extension of the straight-line relationship (Eq. 37), established using only the first six members of a given $Z(CH_2)_nH$ series (Figs. 56–58), is also valid for the liquids of that series with $n > 6$, then one may estimate the effective solubility parameter, δ_n', for those liquids with $n > 6$ on the basis of the corresponding $(\delta_{pol} - \delta_n')^2$ scaled on the abscissa as indicated graphically in Figs. 56–58. The effective solubility parameters (δ_n'), determined in this way, are collected in Table 8. If one assumes further that the difference between δ_n' and corresponding δ_n, calculated on the basis of additive component contribution to cohesive energy, is attributable primarily to self-association [163] then it becomes possible to calculate the corresponding effective molar volume (V_n') of the test-liquid, on the basis that V_n' should equal $E_n/(\delta_n')^2$, where E_n is the cohesive energy of the test liquid. The corresponding fractional increase, X_n, in molar volume owing to self-association is then given by;

$$X_n = (V_n' - V_n)/V_n . \tag{38}$$

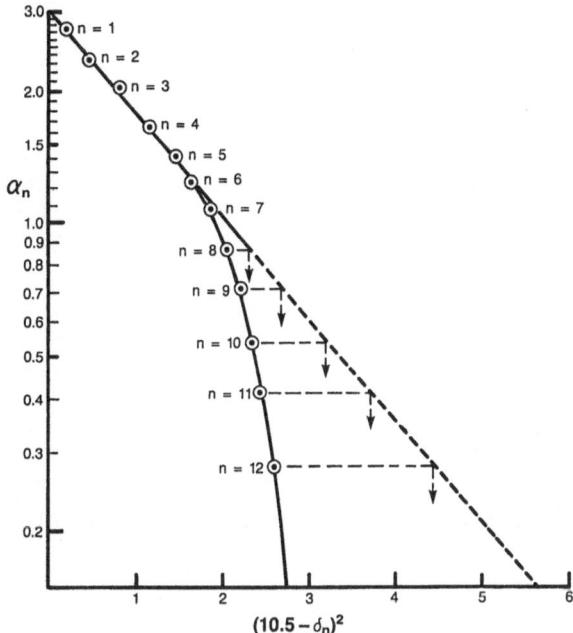

Fig. 58. Correlation of Log α_n for $Br(CH_2)_nH$ with the *calculated* $(10.5 - \delta_n)^2$, showing the corresponding *effective* $(10.5 - \delta_n)^2$ for those liquids with $n > 6$

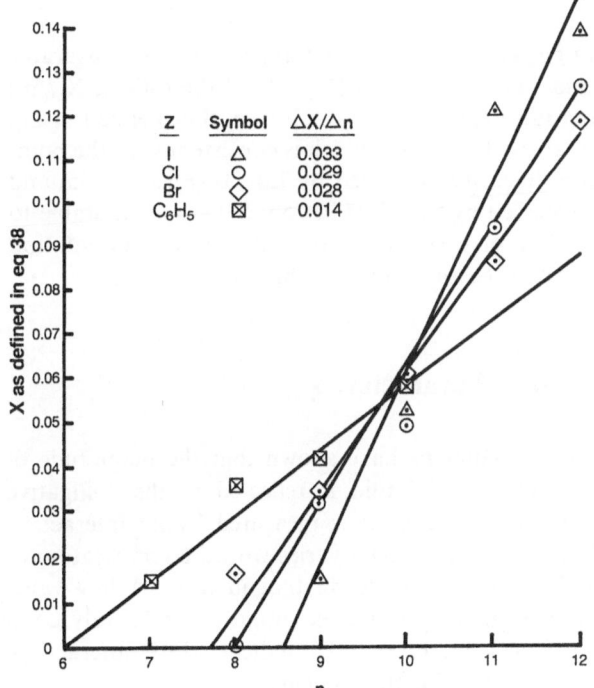

Fig. 59. Correlation of X, the fractional change in molar volume, $(V' - V_n)/V_n$, with n for $Z(CH_2)_nH$ liquids that have $n > 6$

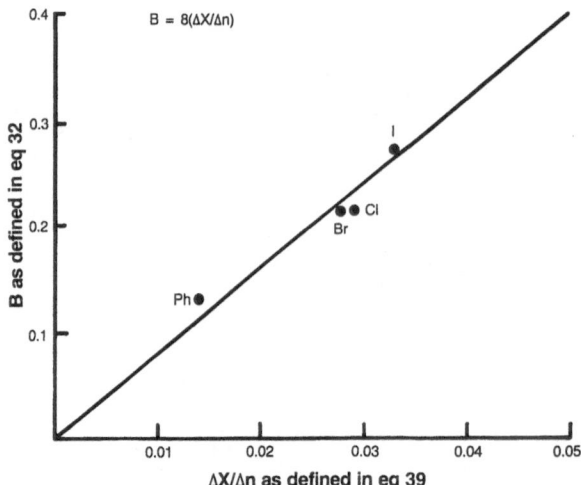

Fig. 60. Correlation of B (Eq. 32) with $\Delta X/\Delta n$ (Eq. 39) for $Z(CH_2)_nH$ liquids, where Z is Ph, I, Br or Cl

The calculated X_n is correlated with n for each $Z(CH_2)_nH$ series in Fig. 59, which shows that X_n increases with the difference $n - n'$, in accordance with Eq. 39:

$$X_n = (\Delta X_n/\Delta n)(n - n'), \tag{39}$$

where n' is the value of n that corresponds to the first appearance of measurable X_n in the given $Z(CH_2)_nH$ series. The correlation (Fig. 60) of the ratio $(\Delta X_n/\Delta n$; Eq. 39) with the corresponding ratio $\Delta \text{Log } \alpha_n/(F_n - F')$, i.e. B as defined in Eq. 32, shows that B increases linearly with $\Delta X_n/\Delta n$, which is consistent with the point of view that the deviation from linearity expressed by Eq. 30 (Figs. 32–35) and the deviation from linearity expressed by Eq. 37 (Figs. 56–58) are attributable to the same cause, namely self-association owing to correlated molecular orientation of London dispersion forces when n becomes greater than 6.

4.2 Flory-Huggins Interaction Parameter, χ

As stated in Sect. 2.2.2, Flory and Huggins have shown that the magnitude of association between polymer and sorbed liquid is reflected in the colligative physical properties of the resultant mixtures. It is measured by an interaction parameter, χ, as defined in Eqs. 10 and 12, which varies with the temperature of the mixture and the volume fraction of polymer in that mixture. A low value ($\chi = <0.3$) indicates considerable interaction, i.e. the liquid is a relatively good solvent for the polymer, and a high value ($\chi = >0.8$) indicates little interaction, i.e. the liquid is a relatively poor solvent for that polymer.

Comprehensive tabulations of χ, reported by numerous investigators for various polymer-liquid systems, have been compiled by Orwoll [43], who classified these data according to polymer (P), liquid (L), volume fraction of polymer (v), and temperature (T). Of the 85 χ-values reported for the P-L systems in which P is polystyrene (Table XIX of Ref. 43), 23 had been measured at 25 °C, and therefore only these data were available for correlation with the swelling powers, C, determined for the corresponding liquids at 23 °C in 3M laboratories [180].

Fortunately additional χ-data can be deduced from the data reported in Table XIX of Ref. 43, based on the assumption that χ at a given T varies linearly with v, which was inferred [180] from the plots of χ at 25 °C vs v for the polyisobutylene-benzene and polystyrene-methyl ethyl ketone systems recorded respectively in Figs. 1 and 2 of Ref. 43. These correlations show that for these two P-L systems χ tends to increase linearly with v from χ_0 = about 0.48 at v = 0 to χ_1 = about 0.85 at v = 1.0.

This assumption is supported by similar correlations [180] of χ with v of P-L systems for which χ-values at two or more levels of v were reported (Tables III to XXII of Ref. 43). Examples of such correlations are recorded here in Fig. 61. The χ-data reported in Table XIX of Ref. 43 for polystyrene-liquid systems at

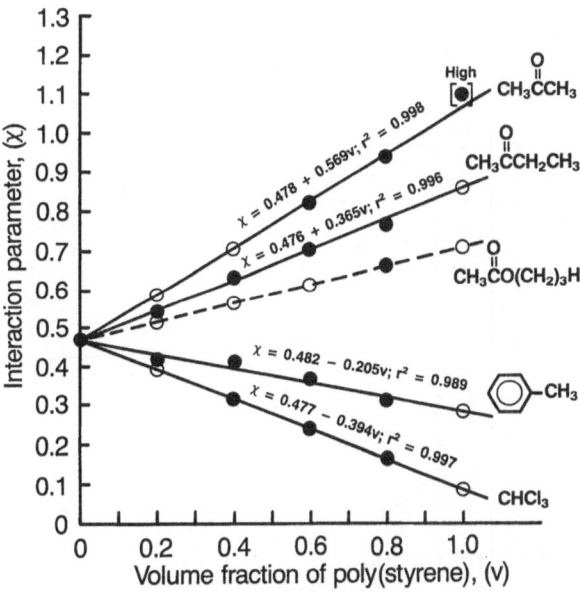

Fig. 61. Correlation of the Flory-Huggins interaction parameter, χ, for polystyrene-liquid systems at 25 °C with the volume fraction (v) of polymer in the system. The *filled circles* represent experimentally determined data recorded in Table XIX of Ref. 43. The *empty circles* represent estimations by interpolation or extrapolation of the linear relationships established on the basis of the experimental data shown. The value for χ reported for acetone at v = 1 is placed in brackets to indicate that this point seems too high, and therefore it was not included in the data set used to establish by linear regression the equation shown for this linear relationship

25 °C in which the liquid is acetone, methyl ethyl ketone, n-propyl acetate, toluene and chloroform, are represented in Fig. 61 by filled circles. The lines of best fit through these five sets of data pass through a common point of intersection at $\chi_0 = 0.48 \pm 0.01$ where v = 0. The linear relationships for the relatively poor solvents (ketones) exhibit positive slopes, whereas those for the relatively good solvents (toluene and chloroform) exhibit negative slopes. Since only one point for n-propyl acetate was usuable, a dashed line was drawn through this point and that at $\chi_0 = 0.48$. The general nature of such relationships, deduced from the data reported in this and the other Tables reported in Ref. 43, encourages one to estimate χ at v levels of the respective P-L systems that were not determined by experiment. Accordingly these missing χ-data for such polystyrene-liquid systems were estimated by interpolation or extrapolation [180], and these data are indicated by the empty circles recorded in Fig. 61.

The union of set of data reported in Table XIX of Ref. 43, and the set of data estimated therefrom, as noted in Fig. 61, were then correlated [180] with the relative swelling powers, C, of the corresponding liquids (Fig. 62) at a given level of v ranging from 0 to 1 in increments of 0.2. The data at each respective level of v is represented in Fig. 62 by a unique symbol as noted in the caption for Fig. 62.

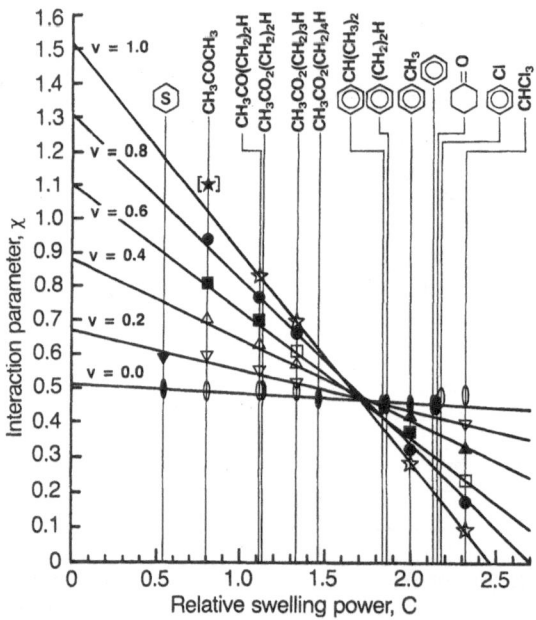

Fig. 62. Correlation of the Flory-Huggins Interaction Parameter, χ, for polystyrene-liquid systems at 25 °C with the relative swelling power (C) of the corresponding liquid at 23 °C as a function of the volume fraction (v) of the polymer in the system. The χ-data at v = 1.0, 0.8, 0.6, 0.4, 0.2, and 0.0 are represented respectively by the symbols: *star, circle, square, triangle pointing upward, triangle pointing downward,* and *narrow oval.* The χ-data determined experimentally are represented by the filled symbols, whereas those obtained by interpolation or by extrapolation as noted in Fig. 61 are represented by the corresponding empty symbols

Those symbols that are filled represent the data determined experimentally and reported in Table XIX of Ref. 43, and those that are empty represent data obtained by extrapolation or interpolation as identified in Fig. 61. The set of six correlations (recorded in Table 19) show [180] that in every case χ at a given level of v decreases linearly with C as expressed by:

$$\chi = \chi_0 - AC. \tag{40}$$

The square of the correlation coefficient (r^2; Table 19) to the line of best fit through each set of 5 or more data points is >0.99 when v is >0.2, 0.965 when v is 0.2

Figure 62 also shows that χ_0 and A, as defined in Eq. 40, both increase with v. The correlations of χ_0 and A with v (Table 19; Fig. 63) show that both relationships are linear as expressed respectively by Eqs. 41 and 42:

$$\chi_0 = 0.49 + 1.01v, \tag{41}$$

$$A = 0.012 + 0.59v. \tag{42}$$

If one assumes that A at v = 0 is zero instead of 0.012 (which Fig. 63 indicates may be too high) and redoes the regression analysis accordingly, r^2 increases to 0.9997, and the relationship of A to v is given by 43 instead of 42.

$$A = 0.610v. \tag{43}$$

Direct substitution of Eqs. 41 and 43 into Eq. 40 gives:

$$\chi_v = 0.49 + 1.01v - 0.61vC. \tag{44}$$

Thus, at v = 1:

$$\chi_1 = 1.50 - 0.610C \tag{45}$$

and on the basis of Fig. 61, χ_v at any other value of v is given by:

$$\chi_v = 0.49 + (\chi_1 - 0.49)v. \tag{46}$$

Table 19. Summary of regression analysis data for $\chi = \chi_0 - AC$

v	0.0	0.2	0.4	0.6	0.8	1.0
χ_0	0.51	0.67	0.80	1.08	1.30	[1.51]
A	0.03	0.12	0.24	0.36	0.49	[0.61]
N	13	7	5	5	5	5
r^2	0.574	0.965	0.993	0.996	0.998	0.996

v is the volume fraction of polymer in the polystyrene-liquid system.
χ_0 and A are as defined in Eq. 40.
N is the number of data points used to establish Eq. 40.
r^2 is the square of the correlation coefficient to the line of best fit (Fig. 62; Eq. 40) through the set of N data points for liquid-polymer systems with a given v.

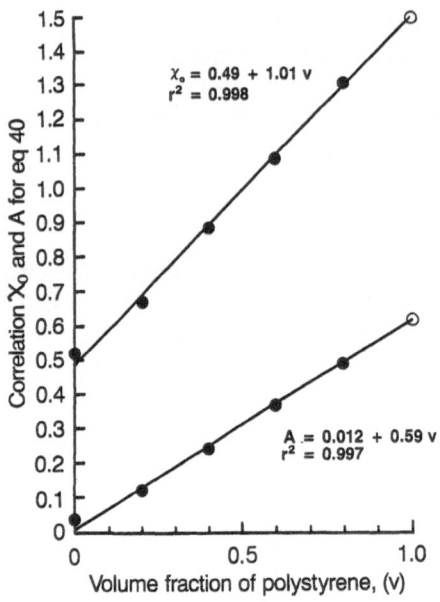

Fig. 63. Correlation of Parameters χ_0 and A, established for polystyrene-liquid systems as a function of C and v (Eq. 40), with v. The *filled circles* represent the data established in Fig. 62 for v ranging from 0.0 to 0.8 in increments of 0.2, and the *empty circles* represent the corresponding estimated data established by extrapolation of the line of best fit through the data represented by the filled circles

The value of χ_1 can thus be determined very accurately on the basis of C using Eq. 44, and considerably less accurately by the volume retention method using chromatography [43]. The values calculated from the values of C determined thus far in ongoing studies of polymer swelling at 3M are reported in Tables 2–17, along with the corresponding δ values for sake of comparison.

Since C is proportional to the product of α, i.e. the number of adsorbed molecules per phenyl group of polystyrene in solution, and the molar volume of the sorbed liquid, i.e. C = (M/d) (α/104), direct substitution into Eq. 44 gives:

$$\chi = 0.49 + 1.01v - 0.61v(M/d)\,(\alpha/104) \tag{47}$$

and at v = 1 Eq. 47 becomes:

$$\chi_1 = 1.50 - 0.00587(M/d)\,\alpha, \tag{47a}$$

where M and d are the formula weight and density respectively of the sorbed liquid, and 104 is the formula weight of a styrene unit. Since α reflects the molecular structures of the adsorbed liquid and the repeat unit of polymer as described in Sect. 3, Eq. 47a states that χ is also a function of the molecular structure, and since the relationships established using poly(Sty-co-DVB) have been shown to hold in a qualitative sense for all polymer liquid systems (see Sect. 3.3.2), it should someday be possible to calculate χ on the basis of the molecular structures of the polymer and solvent. Hopefully this will come about more quickly as a result of the ongoing studies at 3M that are directed toward a fundamental understanding of polymer swelling and drying.

4.3 Guenet Adsorption Parameter, $\bar{\alpha}$

As described earlier (Sect. 2.3) Guenet's study of thermally-induced conversion of isotactic polystyrene solutions to rigid gels led to his observation that the ratio, $\bar{\alpha}$, of residual adsorbed solvent molecules per phenyl group of polymer, after all of the non-adsorbed liquid has been removed by evaporation in vacuum, is characteristic of the liquid [79–82, 181]. This end-point was established by correlating the heat of fusion of the liquid in the system with the amount of residual sorbed liquid. The composition extrapolated to $\Delta H = 0$ was taken to be $\bar{\alpha}$. He later reported [181] that his thermodynamic protocol is also applicable for the determination of adsorbed liquid by atactic polystyrene.

If one accepts that the physical integrity of these gels is attributable to some form of self-association of solvated polymer, as suggested by the earlier investigators, then it follows that $\bar{\alpha}$ must be an averaged value of two polymer fractions, [182], i.e. the fraction (y) that underwent self-association, and the rest $(1 - y)$ that did not. Thus $\bar{\alpha}$ should be expressed by:

$$\bar{\alpha} = (1 - y)\,\alpha_s + y\alpha_a\,, \tag{48}$$

where α_s is the number of adsorbed molecules per repeat unit of polymer that has not undergone self-association (and therefore retains its full complement of adsorbed molecules), and α_a is the number of adsorbed molecules per phenyl group in the polymer that has undergone some form of self-association. In those cases for which α_a is much smaller than α_s, the term $y\alpha_a$ in Eq. 48 may be ignored, and the magnitude of y can then be estimated [182] by the simplified form of Eq. 48, i.e.:

$$y \cong 1 - \bar{\alpha}/\alpha_s = 1 - A \tag{49}$$

which implies that it may be possible to estimate y for a given Polystyrene-Liquid (P-L) system on the basis of the adsorption ratio, $A = \bar{\alpha}/\alpha_s$, for that P-L system.

Accordingly all of the $\bar{\alpha}$-values reported thus far by Guenet for atactic and isotactic P-L systems, along with the corresponding α_s-values (reported here in Tables 2–19), are collected in Table 20 for easy comparison, which shows that A for these P-L systems vary from 2.07 for cyclohexene to 0.31 for p-chlorotoluene. The correlation of $\bar{\alpha}$ with α_s (Fig. 64) shows that for a given polymer tacticity and within a given class of liquids, $\bar{\alpha}$ varies linearly with α_s. The linear relationships for atactic P-L systems with $A < 1$ appear to pass through the origin and therefore may be expressed by the general equation:

$$\bar{\alpha} = A\alpha_s\,, \tag{50}$$

whereas those for isotactic P-L systems appear to intersect the abscissa at $\alpha_s = -0.5$, and thus these may be represented by the general expression:

$$\bar{\alpha}^* = A(\alpha_s + 0.5) = A\alpha_s^*\,. \tag{51}$$

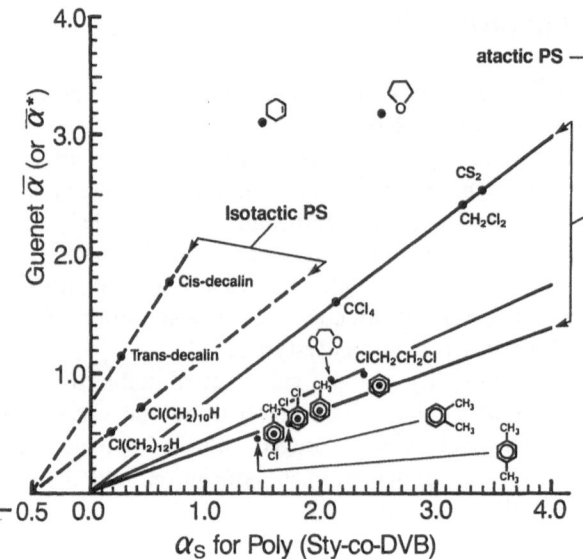

Fig. 64. Correlation of Guenet's $\bar{\alpha}$ (for atactic) and $\bar{\alpha}^*$ (for isotactic polystyrene) with the adsorption parameter α_s determined for the corresponding liquid with respect to poly(Sty-co-DVB)

Table 20. Adsorption parameters for solvated atactic polystyrene (α_s) compared to the corresponding Guenet parameter for atactic ($\bar{\alpha}$) or isotactic polystyrene ($\bar{\alpha}^*$)

No.	Liquid	(Polymer Tacticity)	$\bar{\alpha}$	α_s	α_s^*	A	y
1	THF (a)	atactic	3.22	2.57	—	1.25	0.25
2	cyclohexene	atactic	3.10	1.50	—	2.07	1.0
3	carbon disulfide	atactic	2.55	3.37	—	0.76	0.24
4	methylene chloride	atactic	2.40	3.24	—	0.74	0.26
5	carbon tetrachloride	atactic	1.60	2.12	—	0.75	0.25
6	1,2-dichloroethane	atactic	1.00	2.39	—	0.42	0.58
7	p-dioxane	atactic	0.97	2.10	—	0.46	0.54
8	benzene	atactic	0.92	2.50	—	0.37	0.63
9	toluene	atactic	0.70	1.98	—	0.35	0.65
10	o-dichlorobenzene	atactic	0.64	1.80	—	0.36	0.64
11	o-xylene	atactic	0.59	1.74	—	0.34	0.66
12	p-chlorotoluene	atactic	0.50	1.59	—	0.31	0.69
13	p-xylene	atactic	0.46	1.46	—	0.32	0.68
14	cis-decalin	isotactic	1.75*	0.68	1.18*	1.48*	0.48
15	trans-decalin	isotactic	1.15*	0.28	0.78*	1.47*	0.47
16	1-chlorodecane	isotactic	0.70*	0.44	0.94*	0.74*	0.25
17	1-chlorododecane	isotactic	0.50*	0.18	0.68*	0.74*	0.26

$\alpha_s^* = \alpha_s + 0.5$

A is the association ratio, i.e. $\bar{\alpha}^*/\alpha_s^*$ for isotactic P-L systems and $\bar{\alpha}/\alpha_s$ for atactic P-L systems. It is approximately equal to the slope of the lines represented by Eqs. 50 and 50a.

y is the fraction of solvated polymer that underwent some form of self-association.

(a) Guenet cautioned (private communication) that the $\bar{\alpha}$ measured for THF (Ref. 80) may in fact be a bit high and needs to be repeated.

The displacement in the amount of 0.5 units along the abscissa (Fig. 64) is consistent with Guenet's observation that for a given sorbed liquid, the $\bar{\alpha}^*$ observed for isotactic P-L systems is uniformly greater than $\bar{\alpha}$ observed for the corresponding atactic P-L system. It was inferred from these results that α_s, which was determined using atactic P-L systems, is equal to $(\alpha_s^* - 0.5)$. This implies that the dynamic adsorption density (α_s^*, which was not measured directly) on a monomer unit of isotactic polymer in solution is greater than that (α_s) on a monomer unit of atactic polymer in solution as noted above and as recorded in Table 20.

It appears, therefore, that A may be characteristic of the class of liquids sorbed by a polymer of a given tacticity (as noted in Fig. 64). The data points for atactic P-L systems with liquids that contain only one carbon atom (CS_2, CH_2Cl_2 and CCl_4) fall on a line with slope A equal to about 3/4, those that are aliphatic liquids that contain more than one carbon atom and at least two non-carbon atoms (1,2-dichloroethane and p-dioxane) fall on the line with slope A equal to about 1/2, and those that are aromatic liquids (six in all) fall on the line with slope A equal to about 1/3.

It is believed that the two straight lines that can be drawn (not shown in Fig. 64) from the origin to the data points for cyclohexene ($\bar{\alpha} = 3.10$; $\alpha_s = 1.50$) and for tetrahydrofuran ($\bar{\alpha} = 3.32$; $\alpha_s = 2.57$) may represent the linear relationships (Eq. 50) for atactic P-L systems in which the respective class of liquids would be cyclic aliphatic olefins and cyclic aliphatic ethers.

Figure 64 also shows that the data points for isotactic P-L systems with alpha-substituted linear alkanes (1-chlorodecane and 1-chlorododecane) fall on the line with slope A equal to about 2/3, whereas those for cis-decalin and trans-decalin fall on the line with slope A equal to about 3/2.

It is obvious that the assumption "α_a is much smaller than α_s" can only be possible when the adsorption ratio for a given P-L system is less than 1, and therefore estimation of y on the basis of Eq. 49 is limited to those P-L systems that exhibit slopes A (i.e. $\bar{\alpha}/\alpha_s$ or $\bar{\alpha}^*/\alpha_s^*$) that are less than 1. Thus, for the liquids 1-chlorodecane and 1-chlorododecane (isotactic P-L systems) and the liquids CS_2, CH_2Cl_2 and CCl_4 (atactic P-L systems), y is about 1/4; for 1,2-dichloroethane and p-dioxane, y is about 1/2 and for the six aromatic liquids (atactic P-L systems), y is about 1/3.

It is also obvious from Eq. 48 that y cannot be determined for a P-L system with $A > 1$ unless α_a can be established by some other physical method. Unfortunately this is not yet possible. If one were to assume, however, that the $A = 2.07$ for cyclohexane represents the extreme case, in which y has become equal to 1, then it follows from Eq. 48 that α_a for this P-L system must be equal to $2\alpha_s$. If one postulates further that α_a for the other three P-L systems with $A > 1$ (Table 20) are also equal to about $2\alpha_s$, then Eq. 48 for such systems reduces to:

$$y \cong A - 1 \tag{52}$$

which indicates that y for THF (atactic P-L system) is about 1/4, and y for cis-decalin and trans-decalin (isotactic P-L systems) is about 1/2 (Table 20).

Since α_s was shown to reflect the molecular alignment of the sorbed liquid relative to the monomer unit in "true solution" (see Sect. 3.4) then $\bar{\alpha}$ or $\bar{\alpha}^*$ must also reflect how well the molecular structure of the adsorbed liquid is accommodated by that of the monomer unit of polymer at its respective level of polymer swelling, i.e. after a sizable fraction (y) of that solvated polymer has undergone some form of self-association and all of the non-adsorbed molecules have been eliminated.

In this connection it should be pointed out that there is a formal analogy between $\bar{\alpha}$ and the quantity α'_s, the latter of which is determined kinetically in time-studies that monitor the number, α_t, of residual sorbed molecules per phenyl group of $(Sty)_{1-x}(DVB)_x$ as a P-L system is allowed to evaporate at 23 °C from liquid-saturation to virtual dryness (see Sect. 3.3.2 and Fig. 21). The composition for which all of the molecules not immobilized by adsorption (Fig. 20) have been eliminated is defined as α'_s, and it is indicated by a qualitative change in the kinetics from zero-order to first-order (Fig. 21), which signals incipient elimination of adsorbed molecules. The composition α'_s is related to α_s by Eq. 25, i.e. $\alpha'_s = 0.31 \Lambda(1 + \alpha_s)$.

It is clear, however, that α'_s is not quantitatively equal to $\bar{\alpha}$ because the macrostructural "looseness" (Λ) of the polymer in the two cases is not the same; Λ for $(Sty)_{1-x}(DVB)_x$ is given by $[(1/x)^{1/3} - (1/x)_0^{1/3}]$, as defined in Eq. 20, whereas Λ for the non-crosslinked polymer varies with the class of liquid, which determines y and the distribution of the self-associated domains that comprise y as noted above. It may be possible, however, to establish a quantitative relationship of $\bar{\alpha}$ with α_s for a given class of liquids, which could then be used in turn to establish the corresponding $\bar{\alpha}/\alpha_s$ for other P-L systems in that liquid classification.

The adsorption ratios, $A = \bar{\alpha}/\alpha_s$ or $\bar{\alpha}^*/\alpha_s^*$, observed thus far for atactic and isotactic systems indicate that self-association can occur in two ways; either by expulsion of already-adsorbed molecules (i.e. in the cases that A is <1) or by addition of more adsorbed molecules (i.e. in the cases that A is >1). Why this is so is not understood. It is curious to note, however, that the solvent in those P-L systems with A > 1 is in every case a cyclic aliphatic liquid, the ring structure of which contains no more than one atom that is not carbon (Nos. 1, 2, 14, and 15; Table 20), whereas none of the solvents in a P-L system with A < 1 is in this category. Because the number of $\bar{\alpha}$ reported thus far are relatively few (Table 20), owing to the time-consuming procedure and the high technical skill required to obtain data via the protocol described by Guenet, it is not yet possible to adjudicate with certainty whether or not this cyclic *vs* acyclic differentiation is a real phenomenon or just a fortuitous observation. It does suggest the possibility, however, that the mode of adsorption in the case of cyclic aliphatic molecules may be qualitatively different from that for acyclic molecules.

The above observations are consistent with the comparisons described in Sect. 3.4.2, which show that α_s for cyclic $ZCR(CH_2)_n$ liquids is uniformly greater than that of the corresponding linear $ZCRH(CH_2)_nH$ liquid, and that the magnitude of the difference is markedly greater than that expected on the basis of decreased steric hindrance owing to cyclization of the linear chain attached to substituent Z. This enhanced adsorptivity is not unique to alpha-substituted alkanes; it appears to be general for all classes of liquids (based on the adsorptive functionality of the group Z) as indicated by the data collected in Table 16. Here again the nature of this qualitative difference is not yet fully understood [182]. Obviously many more experiments need to be carried out in order to elucidate fully why the mode of adsorption of cyclic aliphatic compounds appears to be different from that of the linear counterparts.

5 Concluding Remarks

These studies of liquid sorption by poly(Sty-*co*-DVB) particles enmeshed in PTFE microfibers have shown that an adsorption parameter, α, characteristic of the molecular structure of a sorbed liquid with respect to that of the monomer unit in the polymer, can be determined with good precision (± 1 in the third significant figure) by measuring the weight of sorbed liquid per unit weight of enmeshed particles. These adsorption parameters appear to be of fundamental scientific value and therefore potentially useful in understanding natural phenomena in which adsorption plays an important role.

From the beginning of these studies it was clear that the information gained thereby might provide better insight into the mechanism of permeation through polymer films in cases for which the concentration of permeant in the polymer became large enough to cause incipient transition from the glassy or crystalline states of the polymer to the rubbery state. So long as the system remains rigid (i.e. the concentration of the permeant in the system is well below that of the transition composition) the theories expounded by Vrentras and Duda [16], Koros [18], Berens [45] and others [6, 9, 14, 106] (Sect. 2), which in effect are based on permeation of molecules through rigid gratings on a molecular scale, certainly apply; but when the molecular "grating" begins to expand owing to adsorption, and the polymer molecules begin to attain the mobility characteristic of the rubbery state, the mechanism of permeation changes accordingly, and principles based on separation through a rigid molecular "grating" can no longer apply.

At the extreme state of swelling, the physical properties of a liquid-saturated film barrier are that of a gel, i.e. a liquid-supported membrane. In such cases the mechanism of permeation will be dependent largely on the solubility and subsequent diffusion of the permeant through the sorbed liquid. This will be modified by adsorption/desorption properties with respect to the polymer that supports the liquid-saturated membrane. Addition to such systems of a non-volatile solute that has the right affinity for the permeant might perhaps add the dimension of facilitated transport to enhance selectivity at high flux rate. it is hoped, therefore, that the knowledge gained by studying α as a function of molecular structure will improve our understanding of permeation and separation processes involving polymeric membranes over the full range of composition, i.e. from the virtually dry state of the system to the liquid-saturated state.

It is obvious that a quantitative understanding of α as a function of molecular structure of the adsorbed species would be very useful in chromatographic separations. If this data correlates meaningfully with the enormous body of empirical data already accumulated for gas-polymer and liquid-polymer chromato-graphy, then it may be possible to use the latter body of data in turn to calculate α for compounds that are normally solids at about 25 °C and therefore unsuited to being determined by the protocol established for measuring polymer swelling as described in Sect. 3.1. A possible step in this direction is the excellent correlation of C with the Flory-Huggins Interaction Parameter, χ, thus enabling one to establish χ at any level of polymer fraction (v), by use of observed C-values. Chromatographic techniques have also been used to measure χ, and indeed represents the best of the earlier methods for measuring χ at v = 1, where χ is most sensitive to the molecular structure of the sorbed molecule.

It is also obvious that a quantitative understanding of α as a function of molecular structure might help elucidate observed solvent effects upon catalytic reactions and upon the kinetics of reactions in solution. The rate and selectivity of a given reaction must be affected markedly by the "bulkiness" of the molecules immobilized by adsorption to sites close to those mutually attractive electropositive and electronegative centers that are actually involved in certain organic reactions. Thus far the major emphasis [66] has been on acquiring understanding of how the solvent affects the stability of the transition states. The nature of sorbed species, which can serve as "spacers" between the mutually attractive centers, has been largely ignored, despite that such "spacers" can in fact impede severely the attainment of the proximity required to enable the short-range electronic forces to become dominant.

The results summarized in this review represent the early stages of an ongoing research effort at 3M, which in effect is a "three-dimensional approach to surface chemistry", i.e. an area of science between surface chemistry and solution chemistry. The sorption sites on the polymer are not representable as an immobile surface between solid and liquid domains, but rather are string-like polymer molecules undulating in a sea of liquid, so that the adsorption phenomenon must be represented three-dimensionally on a molecular scale rather than two dimensionally.

The strategy of this research will continue to be the determination of α for various homologous series of liquids ZR, in which the functional group in the test-liquid is kept constant and the rest of the molecule is modified systematically. The complexity of such series will increase progressively to enable interpretation of the mode of adsorption by comparison with observation and data accumulated using the relatively uncomplicated series described thus far. Since an understanding of desorption helps in understanding adsorption, studies of evaporation from liquid-swollen polymer systems are being carried out concurrently with those of polymer swelling.

Acknowledgement: I am indebted to Dr. G. V. D. Tiers for reading this manuscript, for many lengthy discussions thereof, and for numerous suggestions for improvement. I am also grateful for his continuous encouragement and, on occasion, active support as indicated specifically in the list of references.

6 References

1. (a) Staudinger H, Heuer H, Huseman E (1935) Trans Faraday Soc 32: 323; (b) Staudinger H, Heuer W (1934) Z Physikal Chem 171A: 129
2. Paul DR (1980) Polym Eng Sci 20: 1
3. Lonsdale HK (1982) J Membrane Sci 10: 81
4. Meares P (1954) J Am Chem Soc 76: 3415
5. Barrer RM, Barrie JA, Slater J (1958) J Pol Sci 27: 177
6. Vieth WR, Howell JM, Hseih JH (1976) J Membrane Sci 1: 117
7. Berens AR, Hopfenberg HB (1982) J Membrane Sci 10: 283
8. Stern SA, Fritsch HL (1981) Ann Rev Mater Sci 11: 523
9. Stern SA, Fritsch HL (1982) Critical reviews in solid state and material science, 2: 123
10. Crank J (1975) The mathematics of diffusion, 2nd edn, Clarendon Press, Oxford
11. Thomas NL, Windel AH (1982) Polymer 23: 529
12. Hui CY, Wu KC, Lasky RC, Kramer EJ (1987) J App Phys 61: 5137
13. Lasky RC, Kramer EJ, Hui CY (1988) Polymer 29: 673
14. Stannet VT (1968) Simple gases. In: Crank J, Park G S (eds), Diffusion in polymers, Academic, New York, chap 2
15. Stannet VT, Koros WJ, Paul DR, Lonsdale HK, Bale RW (1979) Advances in Polymer Science, 32: 71
16. Vrentas JS, Duda JL (1986) Diffusion. In: Kroschwitz JI, Mark HF, Bikales NM, Overberger CG, Menges G (eds) Encyclopedia of polymer science and engineering, vol 5, 2nd edn, Wiley and Son, 36
17. Koros WJ (1988) Membranes and membrane processes. In: McKetta JJ, Cunningham WA (eds) Encyclopedia of chemical chemical processes and design, Marcel Dekker, Inc, New York
18. Koros WJ, Helling MW (1989) Transport properties. In: Encyclopedia of polymer science and engineering, Wiley and Son. Supplement volume, 2nd ed, Wiley and Son, New York, p 724
19. Errede LA, Stoesz JD, Sirvio LM (1986) J App Pol Sci, 31: 2721
20. Flory PJ (1953) Pricinples of polymer chemistry, Cornell University Press, New York
21. Hildebrand JH, Scott RL (1950) Solubility of non-electrolytes, 3rd edn, Reinhold, New York
22. Flory PJ (1941) J Chem Phys, 9: 660, (1942) 10: 51
23. Huggins ML (1941) J Chem Phys, 9: 440
24. Huggins ML (1942) J Phys Chem, 46: 151
25. Huggins ML (1942) Ann NY Acad Sci, 41: 1
26. Huggins ML (1942) J Am Chem Soc, 64: 1712
27. Van Krevelin DW, Hoftyzer PJ (1970) Properties of polymers, Elsevier Scientific, Amsterdam, chap 7
28. Barton AFM (1983) Handbook of solubility parameters and other cohesive parameters, CRC, Boca Raton, FL
29. Fowkes FW, Tischler PO, Wolfe JA, Lannigen LA, Ademu-John CM, Halliwell MJ (1984) J Polym Sci, Polym Chem, ed, 22: 547
30. Drago RS, Vogel GC, Needham TE (1971) J Am Chem Soc, 93: 6014
31. Drago RS, Parr LB, Chamberlain CS (1977) J Am Chem Soc, 99: 3203
32. Hoy KL (1970) J Paint Technol, 42: 76
33. Haggenmacher J (1946) J Am Chem Soc, 68: 1633
34. Hoy KL (1975) Tables of solubility parameters, 3rd edn, Union Carbide Corp, Chemicals and plastics research and development dept, South Charleston, WV
35. Scatchard G (1945) Chem Rev, 44: 7
36. Small PA (1949) J App Chem, 3: 71
37. Magat M (1949) J Chem Phys, 46: 344
38. Scott RL, Magat M (1949) J Poly Sci, 4: 555

39. Boyer RF, Spencer R S (1948) J Poly Sci, 3: 97
40. Bagley EF (1975) Theories of solvency and solution. In: Carver JK, Tess RW (eds), Applied polymer science, ACS Organic Coatings and Plastics chemistry, Washington, DC chap 13
41. Gee G (1946) Trans Faraday Soc, 42: 585, (1946) 42B: 33, (1944) 40: 468, (1942) 38: 418
42. Carpenter DK (1970). In: Mark HF, Gaylord NG, Bikales NF (eds), Encyclopedia of polymer science, Wiley Interscience, New York, vol 12, p 626
43. Orwoll RA (1977) Rubber chemistry and technology, 50: 451
44. Bonner DC (1975) J Macromol Sci Rev, Macromol Chem, C13: 263
45. Berens AR (1989) J Appl Pol Sci, 37: 901
46. Kubo K, Ogino K (1971) Bull Chem Soc Japan, 44: 997
47. Schreiber HP, Tewari YB, Patterson D (1973) J Pol Sci, Pol Phys ed, 11: 15
48. Cruikshank AJB, Winsor ML, Young CL (1966) Proc R Soc, London Ser, A295: 271
49. Patterson D, Tewari YB, Screiber AB, Guillet JE (1971) Macromolecules, 4: 356
50. Leung YK, Eichenger BE (1974) J Phys Chem, 78: 60
51. Brockmeier NF, McCoy RW, Meyer JA (1972) Macromolecules, 5: 130
52. Flory PJ (1949) J Chem Phys, 17: 223
53. Flory PJ, Tatara YI (1975) J Polym Sci, Polym Phys, ed, 13: 683
54. Flory PJ, Rehner J (1943) J Chem Phys, 11: 521
55. Flory PJ (1970) Disc Faraday Soc, 49: 7
56. Sheehan CJ, Biscio AL (1966) Rubber Chem Tech, 39: 149
57. Barton AF (1975) Chem Rev, 75: 731
58. Burrel H (1970). In: Mark HF, Gaylord NG, Bikales NF (eds), Encyclopedia of polymer science, Wiley-Interscience, New York, vol 12, p 618
59. Burrel H (1975). In: Brandrup J, Immergut EH (eds), Polymer handbook, Wiley-Interscience, New York, 2nd edn, p IV-337
60. DusekK, Prins W (1969) Adv Polymer Sci, 6: 1
61. Minnick MG, Shrag JL (1980) Macromolecules, 13: 1690
62. Morris RL, Amelar S, Lodge TP (1988) J Chem Phys, 89: 6523
63. Bastide J (1987). In: Boccara N, Daoud M (eds), Physics of Finely Divided Matter, Springer Verlag, Heidelberg
64. Bastide J, Boue F (1986) Physica, 140A: 251
65. Edwards SF, Vilgis TA (1988) J Phys France, 49: 1635, Akaroni SM, Edward SF (1989) Macromolecules, 22: 3361
66. Reichart C (1979) Solvent effects in organic chemistry, Verlag Chemie, Weinheim
67. Dole M (1949) Ann NY Acad Sci, 51: 705
68. Dole M, Fuller IL (1950) J Am Chem Soc, 72: 414
69. Yokoyama T, Hiraoka K (1980) Polymer bulletin, 2: 183, (1980), 3: 225, (1981), 4: 285
70. Keith HD, Vadimsky RG, Padden FJ (1970) J Pol Sci, Pt A-2, 8: 1687
71. Helms JB, Challa G (1972) J Pol Sci, Pt A-2, 10: 761, (1972), 10: 1447
72. Lemstra PJ, Challa G (1975) J Pol Sci, Pol Phys, ed, 13: 1809
73. Reiss C, Benoit H (1968) J Pol Sci, Pt C, 16: 3079
74. Kobayashi M, Akita K, Tadokoro H (1968) Macromol Chem, 118: 324
75. Giralamo M, Keller A, Miyasaka K, Overbergh NT (1976) J Pol Sci, Pol Phys, ed, 14: 39
76. Sundararajan PR, Tyrer NJ, Bluhm TJ (1982) Macromolecules, 15: 286
77. Wellinghoff S, Shaw J, Baer E (1979) Macromolecules, 12: 932
78. Tan HM, Hilfner A, Moet H, Baer E (1983) Macromolecules, 16: 28
79. Guenet JM, Lotz B, Wittmann JC (1985) Macromolecules, 18: 420
80. Gan JYS, Francois J, Guenet JM (1986) Macromolecules, 19: 173
81. Guenet JM (1986) Macromolecules, 19: 1961, (1987), 20: 2874
82. Guenet JM, McKenna GB (1988) Macromolecules, 21: 1752
83. He X, Hertz J, Guenet JM (1988) Macromolecules, 21: 1757
84. Buckley DJ, Berger M, Poller D (1962) J Polym Sci, 56: 163
85. Wood LA (1976) J Res Nat'l Bur Stand, Sect A, 80: 451
86. Mark JE (1970) J Am Chem Soc, 92: 7252

87. Davrankov, Tsurupa MP (1980) Angewandte makromolekulare chemie, 91: nr 1404, 127
88. Johnson RM, Mark JE (1972) Macromolecules, 5: 41
89. Frolich D, Crawford D, Rozek T, Prins W (1972) Macromolecules, 5: 100
90. Kuhn JH (1967) J Polym Sci, Part C, 16: 859
91. Harrison DJP, Yates WR, Johnson JF (1985). In: Butler GB, O'Driscoll FF, Wilkes GL (eds), Journal of macromolecular science-reviews of macromolecular chemistry and physics, Marcel Dekker, New York, vol C25-No 4, p 481
92. Gleizes J (1937) J Rev Caoutch Plast, 14: 3, (1938), 15: 3
93. Smith MJ, Peppas NA (1985) Polymer, 26: 569
94. Proske RB (1941) Rubber Chem Tech, 14: 489
95. Bobin J (1934) Chim Ind (Paris), 32: 270
96. Garvey Jr BS (1941) ASTM Bull, 109: 19
97. Zhuravlev VA, Yesipov GZ, Biryukova IN, Ushakov GV (1979) Polym Sci USSR, 21: 784
98. Schreiber HP, Holden HW, Barna G (1970) Polym Sci, Part C, 30: 471
99. Chiklis CK, Grasshoff JM (1969) J Polym Sci, Part A-2, 7: 1619
100. Talmon Y, Miller WG (1978) J Colloid Interface Sci, 67: 284
101. Freeman DH (1986) Precise studies of ion-exchange systems using microscopy. In: Marinsky JA (eds), Ion exchange, Marcel Dekker, New York, chap 5
102. Graf J (1977) Sizing with modern image analyzers. In: Stockham JD, Fochtman EG (eds), Particle size analysis, Ann Arbor Science Publishers, Ann Arbor, MI, chap 4
103. Pepper KW, Reichenberg D, Hale DK (1952) J Chem Soc, 3129
104. Roe R, Sherrington DC (1987) Eur Polym J, 23: 195
105. Enscore DJ, Hopfenberg HB, Stannet VT (1977) J Appl Polym Sci, 21: 1795
106. Berens AR, Hopfenberg HB, (1977). In: Li NN (ed), Advances in separation science, vol 3, CRC Press, Cleveland, OH, p 293
107. Enscore DJ, Hopfenberg HB, Stannet VT, Berens AR (1977) Polymer, 18: 1105
108. Hopfenberg HB (1978) J Membrane Sci, 3: 215
109. Berens AR, Hopfenberg HB (1978) Polymer, 19: 489
110. Berens AR, Hopfenberg HB (1978) Polym Sci, Polym Phys, ed, 17: 1757
111. Enscore DJ, Hopfenberg HB, Stannet VT (1980) Polymer Engr Sci, 20: 102
112. Huvard GS, Stannet VT, Koros WJ, Hopfenberg HB (1980) J Membrane Sci, 6: 185
113. Osborn JL, Sarti GC, Koros WJ, Hopfenberg HB (1983) Polymer Engr Sci, 23: 473
114. Connelly RW, McCoy NR, Koros WJ, Hopfenberg HB, Stewart ME (1987) J Appl Polym Sci, 34: 703
115. Stewart ME, Hopfenberg HB, Koros WJ, McCoy NR (1987) J Appl Polym Sci, 34: 721
116. Stewart ME, Sorrells DL, McCoy NR, Koros WJ, Hopfenberg HB (1987) J Appl Polym Sci, 34: 2493
117. Berr-Howell BD, Peppas NK (1985) Polymer, 26: 569
118. Ree BR, Errede LA, Jefson GB, Langager BA (1979) US Pat 4, 153, 661
119. Errede LA, Ronning PM (1980) US Pat 4, 207, 705
120. Errede LA, Stoesz, JD, Upton J (1984) US Pat 4, 460, 642
121. Errede LA, Stoesz JD, Winter GD (1986) US Pat 4, 565, 663
122. Hagen D (1990) Membrane approach to solid phase extraction. In: Pardue HL (ed), A special issue of analytica chemica acta, Elsevier Scientific, Amsterdam
123. Errede LA, Jefson GB, Langager BA, Olson PE, Ree BR, Reichert ME, Sinclair RA, Stofko JJ (1988) Reactive microporous composite membranes. In: Leyden DE, Collins WT (eds), Chemically modified surfaces in science and industry, vol 2: Proceedings of the chemically modified surfaces symposium, Fort Collins, Colorado, June 17–19 1987. Gordon and Breach Science Publishers, New York, p 91
124. Errede LA, Sinclair RA, Newman SJ (1988) Reactive polymers, 8: 201
125. Errede LA (1986) J Appl Polym Sci, 31: 1749
126. Errede LA, Martinucci PD (1980) Ind Eng Chem Prod Res Dev, 19: 573
127. Errede LA, (1984) Coloid and Interface Sci, 100: 414
128. Errede LA (1984) J Memb Sci, 20: 45

129. Private communication from L Cummings, Bio-Rad Corp
130. Errede LA (1986) Macromolecules, 19: 654
131. Doskocilova D, Schneider B, Jakes J (1978) J Magn Resonance, 29: 79
132. Spevacek J, Pusek K (1980) J Polym Sci, Polym Phys, ed, 18: 2027
133. Manatt SL, Horowitz D, Horowitz R, Pinnell RP (1980) Anal Chem, 52: 1529
134. Ford WT, Balakrishnan T (1981) Macromolecules, 14: 284
135. Ford WT, Balakrishnan T (1983). In: Craver CD (ed), Polymer characterization, Advances in Chem, 203: 475
136. Live D, Kent SBH (1982) ACS Symp Ser, 193: 501
137. Errede LA, Newmark RA, Hill JR (1986) Macromolecules, 19: 651
138. Private communications from the respective commercial sources
139. Marchenkov VV, Khitrin AK (1984) Khim Fiz, 3: 1399
140. Szwarc M (1986) Carbanions, living polymers and electron transfer processes, Interscience, New York
141. Rempp P, Herz JE, Borchard W (1978) Model networks. In: Advances in polymer science, 26: Springer-Verlag, New York
142. Beinert G, Belkibir-Mrani A, Herz J, Hild G, Rempp P (1974) Faraday Disc Chem Soc, 57: 27
143. Weiss P, Hild G, Herz J, Rempp P (1970) Makromolecular Chemie, 135: 249
144. Takahashi S (1983) J Appl Polym Sci, 28: 2847
145. Refojo MF (1965) J Appl Polym Sci, 9: 3161
146. Errede LA, Kueker MJ, Tiers GVD, Van Bogart JWC (1988) J Pol Sci, Polym Chem, ed, 26: 3375
147. Errede LA, Van Bogart JWC (1989) J Polym Sci, Polym Chem, ed, 27: 2015
148. Errede LA (1990) J Polym Sci, Polym Chem, ed, 28: 827
149. Errede LA (1990) J Polym Sci, Polym Chem, cd, 28: 857
150. Errede LA, Tiers GVD, Trend JE, Wright BB, J. Polym. Sci.: Polym. Chem. Ed., submitted Jan 91
151. Errede LA, Aus EB, Duerst RW, J. Polym. Sci.: Polym. Chem. Ed., submitted Jan 91
152. Errede LA, Newmark RA, J. Polym. Sci.: Polym. Chem. Ed., submitted Jan 91
153. Loutfy RO (1986) Pure and Applied Chem, 58: No 9, 1239
154. Hayashi R, Tazuke S, Frank CW (1987) Macromolecules, 20: 983
155. Crol SGC (1979) Coating Technology, 51: No 648, 64
156. Gutierrez MH, Ford WT (1986) J Polym Sci, Polym Chem, ed, 24: 655
157. Pochan JM, Beatty CL, Pochan DF (1979) Polymer, 20: 879
158. ten Bricke G, Karasz FE, Ellis TS (1983) Macromolecules, 16: 244
159. Ellis TS, Karasz FE, ten Bricke G (1983) J Appl Polym Sci, 28: 23
160. Errede LA (1986) Macromolecules, 19: 1525
161. Errede LA (1989) J Phys Chem, 93: 2668
162. Errede LA (1990) J Phys Chem, 94: 466
163. Errede LA (1990) J Phys Chem, 94: 3851
164. Errede LA (1990) J Phys Chem, 94: 4338
165. Errede LA (1960) (a) Phys Chem, 64: 1031, ibid (1961), 65: 2262, (b) (1962) J Org Chem, 27: 3425
166. Bauer N, Fajans K, Lewis SZ (1960) Refractometry. In: Weissberger A (ed), Techniques of organic chemistry, physical methods, vol 1, part 2, 3rd edition, Interscience Publishers, New York, chap 18 ps 1139–1281, (b) Fajans K (1947) Chem Eng News, 27: No 13, 900
167. Cines MR (1950). In: Farkas A (ed), Physical chemistry of hydrocarbons, Academic Press, vol 1, New York, chap 8
168. Lundquist M (1965). In: Ekwall P, Groth K, Runnstrom-Reio V (eds), Surface chemistry, proceedings of the second scandinavian symposium on surface chemistry, Academic Press, New York, p 294
169. Lange H, Schwager MJ (1968) Kolloid-Z u Z Polymere, 223: 145
170. Mukerjee P (1970) Kolloid-Z u Z Polymere, 236: 76

171. Orwoll RA, Flory PJ (1967) J Am Chem Soc, 89: 6822
172. Botherel P (1968) J Colloid Sci, 27: 529
173. Tancrede P, Patterson D, Botherel P (1977) J Chem Soc, Faraday Trans 2, 73: 29
174. Fowkes FW (1980) J Phys Chem, 84: 510
175. Errede LA (1991) J. Phys Chem, 95: 1836
176. Hansen CM, Skaarup KJ (1976) Paint technol Mol, 39: 505
177. Errede LA (1986) Macromolecules, 19: 1522, part 5
178. Errede LA (1986) Proceedings, ACS division of polymeric materials science and engineering, 54: 561
179. Krause S (1987) Partial solubility parameter characterization of interpenetrating microphase membranes. In: Lloyd DR (ed), Material science of synthetic membranes, ACS Symposium Series 269, Washington DC, p 351
180. Errede LA, J. Appl. Pol. Sci 1991, in press
181. Klein M, Guenet JM (1989) Macromolecules, 22: 3716
182. Errede LA, Polymer, 1991, in press

Editor: T. Saegusa
Received August 17, 1990

Author Index

Subject Index

Author Index Volumes 1–99